FUTURE NEEDS IN DEEP SUBMERGENCE SCIENCE

Occupied and Unoccupied Vehicles in Basic Ocean Research

Committee on Future Needs in Deep Submergence Science

Ocean Studies Board

Division on Earth and Life Studies

NATIONAL RESEARCH COUNCIL
OF THE NATIONAL ACADEMIES

THE NATIONAL ACADEMIES PRESS
Washington, D.C.
www.nap.edu

THE NATIONAL ACADEMIES PRESS 500 Fifth Street, N.W., Washington, DC 20001

NOTICE: The project that is the subject of this report was approved by the Governing Board of the National Research Council, whose members are drawn from the councils of the National Academy of Sciences, the National Academy of Engineering, and the Institute of Medicine. The members of the committee responsible for the report were chosen for their special competences and with regard for appropriate balance.

This study was supported by Grant No. OCE-0318698 between the National Academy of Sciences and the National Science Foundation, Office of Naval Research, and National Oceanic and Atmospheric Administration. Any opinions, findings, conclusions, or recommendations expressed in this publication are those of the authors and do not necessarily reflect the views of the organizations or agencies that provided support for the project.

Library of Congress Catalog Number 2004101150

International Standard Book Number 0-309-09114-4 (Book)
International Standard Book Number 0-309-52917-4 (PDF)

Cover art shows a compilation of images of deep submergence vehicles. The National Research Council would like to thank the contributors listed below: Department of the Navy, Naval Historical Center provided a schematic of the USS *Monitor* and a line drawing representation of the *H.L. Hunley*; Woods Hole Oceanographic Institution supplied a drawing of *Alvin* as well as an *Alvin* schematic; M. Johnson (committee member), created a line drawing of Oceaneering's remotely operated vehicle, MAGNUM.

Additional copies of this report are available from the National Academies Press, 500 Fifth Street, N.W., Lockbox 285, Washington, DC 20055; (800) 624-6242 or (202) 334-3313 (in the Washington metropolitan area); Internet, http://www.nap.edu

Printed in the United States of America

THE NATIONAL ACADEMIES

Advisers to the Nation on Science, Engineering, and Medicine

The **National Academy of Sciences** is a private, nonprofit, self-perpetuating society of distinguished scholars engaged in scientific and engineering research, dedicated to the furtherance of science and technology and to their use for the general welfare. Upon the authority of the charter granted to it by the Congress in 1863, the Academy has a mandate that requires it to advise the federal government on scientific and technical matters. Dr. Bruce M. Alberts is president of the National Academy of Sciences.

The **National Academy of Engineering** was established in 1964, under the charter of the National Academy of Sciences, as a parallel organization of outstanding engineers. It is autonomous in its administration and in the selection of its members, sharing with the National Academy of Sciences the responsibility for advising the federal government. The National Academy of Engineering also sponsors engineering programs aimed at meeting national needs, encourages education and research, and recognizes the superior achievements of engineers. Dr. Wm. A. Wulf is president of the National Academy of Engineering.

The **Institute of Medicine** was established in 1970 by the National Academy of Sciences to secure the services of eminent members of appropriate professions in the examination of policy matters pertaining to the health of the public. The Institute acts under the responsibility given to the National Academy of Sciences by its congressional charter to be an adviser to the federal government and, upon its own initiative, to identify issues of medical care, research, and education. Dr. Harvey V. Fineberg is president of the Institute of Medicine.

The **National Research Council** was organized by the National Academy of Sciences in 1916 to associate the broad community of science and technology with the Academy's purposes of furthering knowledge and advising the federal government. Functioning in accordance with general policies determined by the Academy, the Council has become the principal operating agency of both the National Academy of Sciences and the National Academy of Engineering in providing services to the government, the public, and the scientific and engineering communities. The Council is administered jointly by both Academies and the Institute of Medicine. Dr. Bruce M. Alberts and Dr. Wm. A. Wulf are chair and vice chair, respectively, of the National Research Council.

www.national-academies.org

COMMITTEE ON FUTURE NEEDS IN DEEP SUBMERGENCE SCIENCE

JOHN A. ARMSTRONG (*Chair*), IBM (retired), Amherst, Massachusetts
KEIR BECKER, University of Miami, Florida
THOMAS EAGAR, Massachusetts Institute of Technology, Cambridge
BRUCE GILMAN, Sonsub, Inc. (retired), Houston, Texas
MARK JOHNSON, British Petroleum, Houston, Texas
MIRIAM KASTNER, Scripps Institution of Oceanography, La Jolla, California
DHUGAL LINDSAY, Japan Marine Science and Technology Center, Yokosuka, Japan
CATHERINE MEVEL, Laboratoire de Géosciences Marines, Paris, France
SHAHRIAR NEGAHDARIPOUR, University of Miami, Florida
SHIRLEY POMPONI, Harbor Branch Oceanographic Institution, Fort Pierce, Florida
BRUCE ROBISON, Monterey Bay Aquarium Research Institute, Moss Landing, California
ANDREW SOLOW, Woods Hole Oceanographic Institution, Woods Hole, Massachusetts
GREG ZACHARIAS, Charles River Analytics, Inc., Cambridge, Massachusetts

Staff

DAN WALKER, Study Director
JOANNE BINTZ, Program Officer
JOHN DANDELSKI, Research Associate
SARAH CAPOTE, Project Assistant

The work of this committee was overseen by the Ocean Studies Board.

Acknowledgments

This report was greatly enhanced by the participants at the first meeting and workshop held as part of this study. The committee would like to acknowledge the efforts of those who gave presentations at the meetings. These talks helped set the stage for fruitful discussions in the closed sessions that followed.

James Bellingham, Monterey Bay Aquarium Research Institute
Peter Brewer, Monterey Bay Aquarium Research Institute
Robert Embley, National Oceanic and Atmospheric Administration
Charles Fisher, Pennsylvania State University
Dan Fornari, Woods Hole Oceanographic Institution
Patty Fryer, University of Hawaii and Deep Submergence Science Committee (DESSC)
Henry Fuchs, University of North Carolina
Dale Green, Benthos, Inc.
Jim Newman, Woods Hole Marine Systems, Inc.
Richard Pittenger, Woods Hole Oceanographic Institution
Tim Shank, Woods Hole Oceanographic Institution
Edith Widder, Harbor Branch Oceanographic Institution
James Yoder, National Science Foundation
Dana Yoerger, Woods Hole Oceanographic Institution
Mark Zumberge, Scripps Institution of Oceanography

The committee is also grateful to a number of people who provided important discussion material and helped to ensure the accuracy of the report: **Peter Auster; Bob Ballard; Stace Beaulieu; Bob Brown; Robert Carney; Robert Collier; members of the Deep Submergence Science Committee; Dolly Dieter; Jennifer Dreyer; Fred Duennebier; Bob Elder; Richard Fiske; Patricia Fryer; Joris Gieskes; R. Grant Gilmore, Jr.; Stephen Hammond; Mat Heintz; James Heirtzler; Taylor Heyl; Jerilyn Hill; Susan Humphris; Jeff Karson; Christopher Kelley; Terry Kerby; John Marr; Marcia McNutt; Anne McGehee Mills; Barbara Moore; David Naar; Frank Parrish; Eric Powell; Andrea Sanico; Andy Shepard; Eli Silver; Craig Smith; Ken Smith; Fred Spiess; David Stein; Robert Steneck; Robert Stern; Bradley Stevens; Shozo Tashiro; Kevin Threadgold; Maurice Tivey; Karen Von Damm; Barry Walden; Megan Ward; Bob Waters; John Wiltshire; Edward Winterer; Beecher Wooding.**

This report has been reviewed in draft form by individuals chosen for their diverse perspectives and technical expertise, in accordance with procedures approved by the National Research Council's Report Review Committee. The purpose of this independent review is to provide candid and critical comments that will assist the institution in making its published report as sound as possible and to ensure that the report meets institutional standards for objectivity, evidence, and responsiveness to the study charge. The review comments and draft manuscript remain confidential to protect the integrity of the deliberative process. We wish to thank the following individuals for their participation in the review of this report:

MILLARD FIREBAUGH, Electric Boat Corporation, Groton, Connecticut
PATRICIA FRYER, University of Hawaii, Manoa
JOHN HEIDELBERG, Institute for Genomic Research, Rockville, Maryland
SUSAN HUMPHRIS, Woods Hole Oceanographic Institution, Woods Hole, Massachusetts
ROBERT KNOX, Scripps Institution of Oceanography; University of California, San Diego
MARCIA MCNUTT, Monterey Bay Aquarium Research Institute, Moss Landing, California
DONALD WALSH, International Maritime, Incorporated, Myrtle Point, Oregon
EDITH WIDDER, Harbor Branch Oceanographic Institution, Fort Pierce, Florida

Although the reviewers listed above provided many constructive comments and suggestions, they were not asked to endorse the conclusions or recommendations nor did they see the final draft of the report before its release. The review of this report was overseen by **RADM John B. Mooney**, U.S. Navy (retired), J. Brad Mooney, Associates, Ltd., appointed by the Divison on Earth and Life Studies, who was responsible for making certain that an independent examination of this report was carried out in accordance with institutional procedures and that all review comments were carefully considered. Responsibility for the final content of this report rests entirely with the authoring committee and the institution.

Preface

The remarkable progress in deep ocean science in the past 40 years has been made possible in part by a set of observational platforms and instruments whose capabilities have continued to improve at a very high rate. Some of these platforms support human observers and pilots; some do not. As in the exploration and scientific investigation of space, to which deep ocean science is occasionally but inappropriately analogized, the role of human-occupied versus remotely controlled or autonomous robotic devices is occasionally the subject of disagreement within the deep ocean science community, even though there is little disagreement that enhanced capabilities are needed.

The National Science Foundation's Division of Ocean Sciences asked the National Academies' Ocean Studies Board to study future needs in deep submergence science, with a charge to report on the scientific opportunities as well as on the mix of assets and facilities required to exploit those opportunities. As usual, the committee included recognized experts in scientific and engineering disciplines relevant to the charge. As chair and as an outsider to this specific field, it is a pleasure to commend the committee, all of whose members were diligent in their engagement with our study, and all of whom, without exception, made significant contributions to this report. I believe all members of the committee have an enlarged and deepened understanding of the future needs of deep submergence science as a result of our study. Other members of the deep ocean science community gave generously of their time and insight as well, both in presentations and discussion at our meetings and in thoughtful contributions to the committee's web site.

Finally, it is a pleasure to recognize the outstanding work of the study director, Dan Walker, who brought to our task both broad knowledge of ocean science and great skill in the conduct of National Research Council studies. He was ably assisted by Joanne Bintz, program officer, John Dandelski, research associate, and Sarah Capote, project assistant.

John A. Armstrong, *Chair*
Committee on Future Needs in Deep Submergence Science

Contents

Executive Summary

Basic research carried out at depths greater than 200m in the ocean over the last 40 years has provided dramatic and unique insights into some of the most compelling scientific questions ever posed. Understanding the nature of planetary processes and the fundamental constraints on the nature and evolution of life has driven scientific inquiry to remote areas of Earth and the solar system. Yet when compared to the vast distances involved in space, the deep ocean lies essentially at "our back door." The remoteness and isolation of deep ocean environments makes this region of inner space a particularly fertile field for scientific inquiry. Fundamental contributions to the understanding of processes responsible for plate tectonics and ocean chemistry, as well as the origins of life and mechanisms for speciation, have been made by scientists working at depth in the ocean. There is reason to believe that further discoveries can be expected if adequate access to the ocean depths can be provided.

In recognition of the significant potential that this research holds and the unique and challenging requirements that work in the deep ocean presents, the National Science Foundation (NSF), the National Oceanic and Atmospheric Administration (NOAA), and the Navy have made a significant commitment to provide operational support for these efforts. Much of this exciting science was made possible through the use of deep-diving submersibles, including the now famous human-occupied vehicle (HOV) *Alvin* first launched nearly 40 years ago. Despite the excellent maintenance that has allowed *Alvin* to make more than 3,600 safe dives and modifications allowing it to take a pilot and two scientists to depths of 4,500 meters, periodic calls for its replacement have occurred. Although significant improvements have also been made in the design and operation of remotely operated vehicles (ROVs), autonomous underwater vehicles (AUVs), and a variety of in situ remote sensing and sampling instruments, much of the ocean water column and seafloor remains beyond the reach of the U.S. scientific community.

As a consequence, NSF, NOAA, and the Navy asked the National Academies' Ocean Studies Board to undertake an examination of current and future needs in support of deep ocean research and to develop recommendations about the nature and number of assets necessary to maintain the health of this field. The statement of task required the committee to evaluate the future directions and facility requirements for deep submergence science and to examine the range of potentially applicable technologies that can support basic research in deep-sea and seafloor areas.

Specifically, the charge to the committee was to assess current and projected capabilities of occupied and unoccupied vehicles; make recommendations regarding the mix of vehicles needed to continue to carry out world-class deep submergence science; and discuss innovative design concepts and technological advances that should be incorporated into any new vehicles to support future research needs.

The primary sponsor, the Ocean Sciences Division of the National Science Foundation (NSF/OCE), established that the total construction costs (including related costs such as modifications of support ships) for any new assets should not exceed the upper practical limit for the Division of Ocean Sciences' midsized infrastructure fund (roughly 10 percent of the division's annual budget or approximately $25 million) and not require more than a modest increase beyond inflation in the operating budgets of the supporting agencies for deep submergence vehicles. If multiple types of vehicles were deemed necessary by the committee, suggestions were to be made concerning the optimal phasing of implementation over several years.[1] Cost estimates were to be supplied by the National Deep Submergence Facility (NDSF;[2] a national facility that includes *Alvin, Jason,* and

[1]The submersibles used to support deep ocean research are similar to those discussed in three recent National Research Council reports; *Undersea Vehicles and National Needs* (1996), *Enabling Ocean Research in the 21st Century: Implementation of a Network of Ocean Observatories* (2003a), and *Exploration of the Seas: Voyage into the Unknown* (2003b). *Undersea Vehicles and National Needs* provided a broad vision of the nation's needs for undersea vehicles, while the more recent studies focus on needs to support specific deep-sea activities. *Enabling Ocean Research in the 21st Century* and *Exploration of the Seas* each propose a suite of vehicles to support those endeavors. The recommendations in this report are above and beyond any capabilities called for in those two reports.

[2] In 1974, the "significance of maintaining a core deep submergence operational team was recognized by ONR [Office of Naval Research], NOAA and NSF. . . when they established the National Deep Submergence Facility (NDSF) at the Woods Hole Oceanographic Institution and formulated a Memorandum of Agreement (MOA) to share the operating costs of the facility."

other deep ocean research vehicles) as well as other submersible owners and operators, or outside companies if needed. More refined cost estimates would require a more specific design than is currently available. NSF has indicated that it simply wants some assurance that options explored could reasonably be expected to be completed for $25 million.

ENHANCING THE VALUE OF EXISTING ASSETS

The scientific demand for deep diving vehicles (both human-occupied vehicles and remotely operated or autonomous vehicles) is, at present, not being adequately met. Part of this problem can be traced to the inadequacy of the number and capabilities of existing assets to perform the type of scientific effort associated with deep submergence science funded through NSF and NOAA. This report makes specific suggestions for additions to that asset pool. The way in which existing assets are managed and utilized also contributes to the problem. In this context, management refers to the acquisition, maintenance, and access management of vehicles used to carry out basic ocean research and exploration. In particular, the current management system does not always ensure a match between the requirements of federally funded projects and the appropriate deep submergence assets. Modification of existing assets and construction of new assets to alleviate this mismatch would represent both a significant capital investment and an additional demand for operating funds. Decisions to commit these resources should be accompanied by a commitment to ensure the best use of the nation's deep submergence assets. **The management of the nation's deep submergence assets should, therefore, be clarified and revised to ensure the optimal use of both existing and potential assets in future scientific research.**

There appear to be situations in which deep submergence scientific goals cannot be met by NDSF assets, but can be met by non-NDSF assets. For example, limitations on the viewing capability of *Alvin* and on its capability to achieve neutral buoyancy at multiple times during a single dive make it less suited for certain types of midwater research than some vehicles that are not part of the NDSF asset pool. For this reason, arguments favoring the full utilization of NDSF assets for fiscal reasons have the unintended consequence of restricting the scope of deep submergence science.

A reasonable solution to this problem is to upgrade the capabilities of NDSF assets so that they can be used in all fields of deep submergence science. This would expand the scope of deep submergence science and maintain costs at a reasonable scale. Although this is clearly an important part of the solution, by itself, it may be inadequate. First, these upgrades, if they occur, will not be completed for two to three years; thus some short-term measures are needed. Perhaps more importantly, there is a danger that the existing pattern of use of NDSF assets will simply persist. One way to address both of these problems is for NSF/OCE to provide modest, but immediate, funding to support the use of non-NDSF assets (NOAA currently funds the use of non-NDSF facilities on a modest basis). This funding should not be drawn from the NSF/OCE science program budgets but should be allocated by the NSF/OCE Integrative Facilities Program. An additional benefit of establishing such a fund is that it would provide a gauge of demand for capabilities not provided by current NDSF assets. If, as additional assets become available (either through purchase or construction), the demand for non-NDSF vehicles declines (or never materializes), these funds could be used to address other (non-deep submergence) marine operational needs as determined by NSF/OCE.

Recommendation: NSF/OCE should establish a small pool of additional funds (on the order of 10 percent of the annual budget for NDSF) that could be targeted specifically to support the use of non-NDSF vehicles for high-quality, funded research, when legitimate barriers to the use of NDSF assets (as opposed to personal preference) can be demonstrated.

DEVELOPING NEW ASSETS

Reforming the asset management system to allow for wider (though still limited) access to non-NDSF assets will, by itself, not be sufficient to meet the needs of basic ocean research. Existing assets are simply too limited in their capabilities and capacity, especially at depths greater than 3,000m, to support the growing demand to conduct research over the necessary geographic and depth range. High demand for existing deep-diving assets within the NDSF pool has forced asset managers to place a heavy premium on maximizing operational days while minimizing days in transit. The pressure that this geographic restriction has led to can only be expected to increase as ongoing efforts to address a more scientifically diverse set of problems increase the demand for deep-diving vehicles to work in diverse settings.

Recommendation: NSF/OCE should construct an additional scientific ROV system dedicated to expeditionary research,[3] to broaden the use of deep submergence tools in terms of the number of users, the diversity of research areas, and the geographical range of research activities.

[3]As opposed to those needed to support the Ocean Observatories Initiative.

Furthermore, while such an ROV system can be constructed using well-established subsystems, several factors should be considered during its design. Probably most important overall is the incorporation of several attributes that will greatly enhance its utility as well as its ability to complement existing assets. Some of the most significant attributes include standardized tooling suites, open software and hardware architectures, electronic thruster systems, tether management systems, improved handling systems, camera and lighting systems, and a variable ballast system. The total cost of this system would be approximately $5 million, and it could be built and ready for service within one year of authorization. Using the current University-National Oceanographic Laboratory System (UNOLS) model for marine operations, this ROV system could be mobilized onto the current fleet without any additional hardware. The operational requirements would be roughly equal to those for the current *Jason II*. The operational costs represented by this new ROV should therefore be similar to those of *Jason II* and, thus, would represent a 20 percent increase in the overall operating costs of NDSF. This increase should have a modest impact if it is anticipated and the overall budget is increased incrementally in preparation for the construction and operation of a new ROV. One justification for adding a new ROV system to the NDSF asset pool is to provide even greater geographic range to the growing number of ocean scientists seeking access to deep submergence assets.

Recommendation: NSF/OCE should, after a proper analysis of the cost-benefits of distributed facilities, strongly consider basing this new ROV system at a second location that would minimize the transit time for periodic overhaul and refit of both ROV systems.

The best approach to deep submergence science is the use of a combination of tools, at different scales. Surveys of the seafloor are best achieved using tethered vehicles and AUVs. Experiments and observatory work that require longer time at already well-characterized sites on the seafloor are best conducted with ROVs. Moreover, work at depths greater than 6,500m will definitely require unoccupied vehicles, as long as the expense and risk of constructing and operating HOVs capable of work at these great depths discourage their use. As discussed in Chapter 3, human presence at depth remains a significant lynchpin in the nation's oceanographic research effort. Detailed descriptions of specific sites or work in the water column benefit from the direct observation allowed by HOVs. Despite rapid and impressive growth in the capabilities of unoccupied vehicles (both remotely operated and autonomous), scientific demand for HOV access can be expected to remain high. The capabilities of the existing *Alvin,* however, limit its scientific usefulness for some types of deep ocean research. Improving these capabilities, even without extending its depth range, is clearly necessary if many of the high-priority scientific goals discussed in Chapter 2 are to be achieved.

Recommendation: NSF/OCE should construct a new, more capable HOV (with improved visibility, neutral buoyancy capability, increased payload, extended time at working depth, and other design features discussed in Chapter 4).

The bulk of existing *Alvin* use is at depths considerably shallower than its 4,500-m limit. Even at these shallower depths, scientific demand remains unmet. At the same time, certain scientific goals would be furthered by the acquisition of an HOV with a 6,500-m-depth range. Moreover, under current safety rules, *Alvin* and other HOVs are prohibited from operating in waters deeper than the rated working depth, even if this operation is in the water column. Although various options for extending the safety range should be explored (e.g., a depth-triggered system that would make it impossible for the vehicle to descend below a given depth), it can be assumed that this restriction will remain in place for the foreseeable future. For this reason alone, providing access to an HOV with a greater depth capability would allow its use over a broader geographical range, improving its utility for a portion of the potential user community. As discussed at length in Chapter 4 however, it is not clear at present that a suitable sphere can be obtained to allow the fabrication of the main body of a deeper-diving HOV, especially given the limited funds available to NSF/OCE in the next two fiscal years.

The most promising approaches for moving ahead during the time frame articulated by NSF/OCE would make use of one of two existing spheres. The first is an unused sphere from the Russian *Mir* HOV series (referred to as the Lokomo sphere), which has been rated to 6,000m. The potential for obtaining this sphere must be evaluated by NSF and NDSF. The other available sphere is the titanium sphere used in the existing *Alvin,* which is rated to 4,500m. Although other approaches based on fabricating an entirely new sphere warrant investigation, there is insufficient information at this time to determine the ultimate availability and cost of specifically fabricated HOV spheres. Given the technical and cost uncertainties, and that the scientific justification for conducting HOV operations at depths greater than 4,500m appears to be incremental (i.e., it represents promising but logical extensions of work supported at shallower depths), it is not clear that significant additional resources (i.e., in excess of those needed to fully upgrade the current NDSF HOV capability; as discussed in Chapter 4) should be expended on a new HOV with greatly extended depth capability if that expenditure were to preclude construction of the ROV system recommended in Chapter 4 of this report.

Recommendation: Thus, constructing an HOV capable of operating at significantly greater depths (6,000 meters plus) should be undertaken only if additional design studies demonstrate that this capability can be delivered for a relatively small increase in cost and risk.

To implement these recommendations, NSF and other NDSF sponsors would have to increase funding at a rate of 10-15 percent each year over the next three years to (1) cover the cost of non-NDSF vehicle use and (2) cover the cost of the new ROV. In order to provide the capabilities and capacity to meet existing and anticipated demands, NSF and other NDSF sponsors should take a three-step approach: (1) set aside additional funds called for to support non-NDSF vehicle use as quickly as possible; (2) initiate acquisition of the new ROV in 2004 or 2005; and (3) undertake a detailed engineering study to evaluate the various HOV enhancement options called for in Chapter 4 with an aim of delivering these new platforms by 2006. It is entirely possible, perhaps even probable, that given more time and significantly greater funds, the federal agencies that fund deep submergence research could build a number of platforms with greater capabilities than described here. The statement of task was, however, specifically crafted to ensure that advice provided in this report was appropriate given the current fiscal and programmatic realities facing federal science agencies. If, in the future, these requirements were to change significantly, then the appropriate mix of assets needed to support deep submergence effort should be revisited.[4]

[4]The purpose of this study is to provide NSF with recommendations for consideration regarding activities to provide infrastructure support for basic research at depth in the oceans through NDSF or other means. As such, the discussions in this report are designed to inform this question and are not intended to provide an exhaustive account of all research-related activities carried out at depth or a complete account of all the potential assets that exist. The discussion of assets in this report is limited therefore to those that establish whether adequate deep submergence vehicles exist within or outside the National Deep Submergence Facility. Again, any recommendations made in this report are above and beyond the needs for other large programs such as NSF's Ocean Observatories Initiative or activities falling within the realm of ocean exploration.

1

Introduction

HIGHLIGHTS

This chapter

- Introduces deep submergence science as a specific subset of ocean science (to establish its importance as well as the vagueness of the current definition of deep submergence science)
- Describes the nature of assets used (to demonstrate the variety and distribution of platforms available; i.e., the "mix" of technology)
- Discusses the nature, role, and organization of the National Deep Submergence Facility (to establish how its "management" can foster or limit scientific inquiry)
- Introduces the problem as described by the statement of task (to clarify the breadth of the task)
- Describes the organization of the report (to demonstrate that a logical approach was used and to foreshadow the committee's conclusions)

Major advances in the understanding of the oceans have been achieved in the last 40 years. Thanks to a combination of careful planning and serendipity, ocean scientists have revolutionized the view of life on Earth, changed understanding of global tectonic processes and the role of the oceans in climate change, uncovered lost relics of human history, and discovered hundreds of new species.

Much of this exciting science was made possible through the use of deep-diving submersibles (see Box 1-1, Plate 1a,b), including the well-known human-operated vehicle (HOV) *Alvin,* first launched nearly 40 years ago. Despite the excellent maintenance that has allowed *Alvin* to make more than 3,600 safe dives and modifications allowing it to take a pilot and two scientists to depths of 4,500m, periodic calls for its replacement have occurred. These calls have been prompted by a part of the ocean science community that would like more capable vehicles, defined variously, but including one with better visibility; faster transit time to and from the surface, which would result in increased bottom time; and greater depth capabilities (Brown et al., 2000). Although significant improvements have also been made in the design and operation of remotely operated vehicles (ROVs), autonomous underwater vehicles (AUVs), and a variety of in situ remote sensing and sampling instruments, the deepest part of the water column and bottom remains just beyond the reach of science.

The United States has been a dominant player in ocean sciences for at least 50 years. Since World War II, and throughout the Cold War, the Office of Naval Research (ONR) has been a major driver and funder of this research. ONR financed the construction of much of the equipment and instrumentation required, including deep-diving vehicles such as *Alvin,* which is owned by the U.S. Navy but operated by the National Deep Submergence Facility (NDSF) as a civilian research asset.

The United States is not the only nation active in the deep ocean. More than 200 human-occupied deep submergence vehicles (DSVs)[1] have been built worldwide since World War II, with only a few of them dedicated to scientific research. Japan, France, and Russia all operate their own deep-diving research HOVs (the French *Nautile,* which descends to 6,000m; the Japanese *Shinkai 6500,* going to depths of 6,500m; and the Russian *Mir I* and *Mir II,* which are capable of reaching 6,000m). In addition, the U.S. Navy's *Sea Cliff* submersible replaced *Trieste II* in 1982 and was the first 6,000-m non-bathyscaph HOV. Although several U.S. entities operate submersibles at depths up to 1,000m, very few can exceed that depth and only the *Alvin* can dive below 2,000m. The focus of the U.S. Navy has shifted away from deep water over the last 10 years in response to geopolitical developments that call for greater focus on littoral environments.

---

[1]The term "DSV" has traditionally been used to signify a human-occupied deep submersible. As the nature of these assets has diversified however, other more descriptive terms have been employed to define specific submersible types (e.g., occupied-unoccupied, remotely operated, tethered-untethered). The currently accepted definition of DSV, and the one used in this report, connotes any deep submergence vehicle whether it is human occupied, remotely controlled, or autonomous in its operation.

BOX 1-1
Two Milestones in Deep Submergence Vehicles

In 1934, William Beebe and Otis Barton crouched inside the 54-inch-diameter steel *Bathysphere* and were lowered by an attached cable to a record depth of 922m. With only two small windows and a 250-W light to illuminate the area directly outside the sphere, Beebe and Barton were able to view sea life that had never been seen before in its natural environment. The bathysphere had no propulsion capability and was therefore able to move only vertically in the water column, dependent on the attached cable that was raised and lowered by the mother ship.

Almost 30 years later, Lt. Don Walsh and Jacques Piccard descended in the deepest-diving HOV ever built. In January 1960, the *Trieste* settled onto the bottom of Challenger Deep in the Mariana Trench at 10,915m, the deepest part of the ocean. The *Trieste* was a bathyscaph purchased by the U.S. Navy in 1957 to support naval oceanographic research and was capable of submerging under her own power with limited forward control from a propeller and rudder. Unlike Beebe's bathysphere, the *Trieste* was not dependent on a surface ship for submerged operations. The maximum design depth upon delivery to the Navy was 6,300m, but in 1959 the Navy refitted her with a stronger sphere (rated to 15,244m) and larger buoyancy chambers. Because of the immense pressures at depth (~16,000 pounds per square inch at 10,910m), air could not be forced into ballast chambers to increase buoyancy, so aviation gasoline filled the ballast tanks and heavy iron shot was released when it was time to surface. The *Trieste* made only one trip to the very bottom of the ocean and humans have not returned since then.

In 1963, the *Trieste* researched the debris field of the lost submarine *Thresher*, off the coast of Cape Cod in 2,500m of water. In 1964, the *Trieste II* located and inspected the major *Thresher* hull section. Also, from 1968 to 1970, the modified *Trieste II* surveyed the wreckage of another missing U.S. Navy nuclear submarine, the *Scorpion*, which was found near the Azores in 3,488m of water. The *Trieste II DSV-1* is now retired, but her achievements are legendary.

Thus, ONR has reduced its focus on, and support for, deep submergence science.

DEEP SUBMERGENCE SCIENCE

Deep submergence science is defined both scientifically and operationally. Based on scientific criteria, the deep sea is defined as beginning at 150 to 200m (or the lower limit of the epipelagic zone) (Marshall, 1979; Herring, 2002). The operational definition of deep submergence has been arbitrarily set at depths greater than 1,500 to 2,000m, based primarily on the depth capabilities of *Alvin* and *Jason II*. As a result of this operational definition and funding history, *Alvin, Jason II,* and other assets that are part of NDSF (see Plate 2) are the only submersibles in the University-National Oceanographic Laboratory System (UNOLS) and thus eligible for National Science Foundation (NSF) support from operations funds. NSF's Ocean Sciences Division (OCE) research funding supports projects in deep submergence science at depths shallower than 1,500m (e.g., some of the RIDGE program research, deep-sea larval biology, gas hydrate research, and studies conducted in midwater environments), for which both NDSF and non-NDSF platforms are used. Several HOVs and ROVs are available and appropriate for work at depths shallower than 1,500m, but mechanisms for funding these assets as part of NSF/OCE research proposals are perceived to discriminate against their use (UNOLS, 1999).

Because deep submergence science is often conducted at depths much shallower than 1,500m, the committee has adopted the scientific definition of deep sea (i.e., the area of the ocean greater than 200m) as the basis for its recommendations concerning future needs in deep submergence science. Plate 3 depicts the depths of the oceans' basins and graphically represents the areas in which *Alvin* is capable of diving. Chapter 2 documents the diverse nature and significance of deep submergence science and discusses the geographic and water column depth ranges in which this science has to be conducted.

Techniques for sampling the ocean's depths have evolved over the last century and have generally involved sending a sampling instrument to a point within the water column or to the bottom of the ocean and then retrieving it. Charles Darwin on the *Beagle*, using a simple dredge lowered by a hand line, was one of the first to systematically collect samples from deepwater benthic communities. Today there are many ways of collecting samples from the water column as well as the bottom of the ocean, but these methods all have their limitations. Nets and dredges are apt to damage specimens, especially if they are gelatinous (e.g., jellyfish), and provide little to no context of the surrounding area from which the sample was taken. Additionally, this sampling process is difficult if not impossible to carry out during special events (e.g., hydrothermal vent activity) or situations that require the researcher to exercise extraordinarily fine control (e.g., capturing a fish; inserting a water chemistry sensor into the outflow from a hydrothermal vent—if the probe is off by as much as 1 cm, the data will not be valid). This human dimension of submersible control—which includes, for example, sensory-motor skills, reflexes, proprioception (the sensory feedback often referred to as "muscle memory" for specific tasks), and the pilot's and scientist's ability to develop a cognitive map of the area in which they are working—is applicable to both HOVs and ROVs. Piloting either type of vehicle demands intense concentration and the ability to exert fine control over the vehicle to obtain exact and specific data and samples.

The scientific need to visit the deep ocean, obtain intact samples (in contrast to trawls, for example, which most often damage or destroy specimens), conduct experiments, and view a location in real time has spurred the development of deep submergence vehicles. Whether visiting these depths in person in an HOV or remotely with an ROV, there is a clear and legitimate scientific imperative to continue to develop those technologies that will allow visitation of the deep ocean.

Deep submergence science requires a level of sophistication and a technology that can withstand the immense pressures found as ocean depth increases. For humans to directly investigate depths below 300m requires the use of a 1-atmosphere (atm) chamber (i.e., the pressure within the chamber is 1 atm, or that found at sea level), made from glass, steel, massive cast acrylics, or titanium. These are currently the only materials strong enough to withstand the crushing weight of the sea and allow a human to pilot the vehicle with no special protection other than the vehicle itself. Remotely operated and autonomous vehicles do not require any special chamber to protect a human occupant and thus are not constrained by any life-support systems, which make them less expensive to build and certify. Conversely, an HOV has a person on the scene at depth as opposed to a relatively inexpensive ROV equipped with sensors and cameras.

The choices are expensive life-support systems to protect a person viewing the scene through a viewport, versus less expensive remote vehicles that have image capture systems with a human operator on the surface. Although this study does not attempt to answer the question of which system is better in all situations, the strengths and limitations of HOVs, ROVs, and AUVs are considered. It can be said, however, that although ROVs and AUVs could undoubtedly become more sophisticated, possibly supplanting the need for human scientists to directly carry out deep ocean research in many instances, the added value of human perspectives will remain significant.

The National Deep Submergence Facility

The Woods Hole Oceanographic Institution (WHOI) has operated the primary U.S. deep water research HOV, *Alvin*, since 1964, initially supported by a mix of short-term research and engineering contracts and grants. In 1974, the "significance of maintaining a core deep submergence operational team was recognized by ONR, NOAA [the National Oceanic and Atmospheric Administration] and NSF . . . when they established the National Deep Submergence Facility (NDSF) at WHOI and formulated a Memorandum of Agreement (MOA) to share the operating costs of the facility" (UNOLS, 1994). This agreement was later revised to provide a safety net of minimum facility funding to maintain the core capability if research funding for use of the capability dropped too low.

Three years before the formation of the NDSF, UNOLS was established, with assistance from NSF and ONR, to coordinate U.S. oceanographic research ship schedules and facilities. The 1972 UNOLS charter includes provisions for national oceanographic facilities, of which NDSF is a prime example. Consequently, a standing committee of UNOLS, the DEep Submergence Science Committee (DESSC) currently has primary science community advisory responsibilities for the NDSF. UNOLS also coordinates the scheduling of support ships required for submergence operations. In sum, the NDSF was initially formed by MOA among the three primary deep submergence funding agencies, and it also has clear formal links within UNOLS. The vast majority of NDSF vehicle time is funded by NSF/OCE, NOAA's National Undersea Research Program (NURP), and activities of NOAA's Office of Ocean Exploration. Among these programs, NSF/OCE accounts for nearly 80 percent of NDSF vehicle operation days.

Deep Submergence Vehicles

Human-Occupied Vehicles

The *Alvin*, NDSF's only operating HOV, is a three-person vehicle (one pilot and two scientific observers) capable of diving to 4,500m and remaining submerged for 10 hours under normal conditions and up to 72 hours on emergency power. The typical dive profile for *Alvin* is to leave the sea surface by allowing water to enter the main ballast tanks and literally sink under her own weight to the desired depth, usually the ocean bottom. She carries steel plates that make her negatively buoyant for the descent, and some of these are released when the desired dive depth is reached, resulting in neutral buoyancy. The remaining plates are carried throughout the dive and are dropped to obtain positive buoyancy for return to the surface. Once *Alvin* achieves neutral buoyancy at a desired depth, the variable ballast system allows adjustment of the vehicle's weight by plus or minus 250 pounds for vertical excursions between different operating depths, usually limited to a 1,000-m-depth range due to time and battery power constraints. Observations are made through three small viewports (one in front and one on each side), as well as with a number of video cameras coupled to multiple in-hull monitors. The *Alvin* also has a viewport on the bottom that is no longer used for viewing. However, it is equipped with a sensor to ensure that the acrylic port material does not melt when maneuvering around hot vents. In addition to a pair of manipulator arms, various scientific payloads may be attached to the front of *Alvin* for collecting samples and performing experiments. The greatest drawbacks to the *Alvin* are (1) the visibility is limited (viewports are small and not optimally placed for many viewing requirements); (2) the sphere is cramped (observers and pilots must squeeze into awkward positions in order to share available space not occupied by internal system components); and (3) the lead-acid battery energy source can result in power-related dive limitations.

Remotely Operated Vehicles

ROVs are unoccupied, tethered submersibles with an umbilical cable that runs from the pilot (either onboard a mother ship, on land, or even on an HOV) to the ROV. In the United States, the Navy first developed this technology. Major commercial use of ROVs began with the development of the North Sea offshore oil and gas industry in the mid-1970s.

The umbilical cable carries power, pilot control input, and feedback from sensors and video cameras. Because a wireless signal quickly fades, reflects, and is otherwise attenuated under water, the only reliable means of accurately controlling a remote underwater vehicle is through an umbilical cable. While increased bandwidth in acoustic communications and improved task-level control have made remote control of untethered remote vehicles a possibility in the future (NRC, 1996), this technology will likely not include video feed, and thus not provide high-level feedback, reducing the applications for wireless control. Typically, an ROV will have cameras, video transmitted live to the pilot to aid in navigation, high-intensity lights, thrusters for control, manipulators, and a basket or platform for mounting equipment. The operation and control room accommodates several scientists that view the images and interact, in real time, with each other and with the pilot. While most ROVs have a manipulator arm, their functionality can vary dramatically. Some of the drawbacks of ROVs center around the problems associated with operating the vehicle remotely: (1) a cognitive mapping difficulty caused by lack of on-scene navigation; (2) the tether, which can hamper operation, especially in trenches, on walls, and in areas where entanglement may occur; (3) the need for higher-definition cameras and three-dimensional feedback that would attempt to mimic navigation in an HOV; and (4) the lack of a visual feedback mechanism comparable to the human eye, which can make precision piloting extremely difficult (e.g., navigating in reference to pycnoclines that can be seen only as a "shimmer" in the water). Furthermore, tether movement can cause turbulence, which can set off bioluminescent displays that prohibit accurate characterization of the in situ light field. Finally, multiple organisms or particles in three-dimensional space cannot be identified or quantified while either the water mass or the DSV is moving because of focusing and pan-tilt limitations not imposed upon the human eye. Conversely, ROVs offer many benefits including reduced risk to human operators, enhanced potential for collaboration through real-time sharing of information with surface ship and scientists ashore, and virtually limitless bottom time. The best use of the vehicle, however, may differ from mission to mission depending on the needs of the principal investigator.

Autonomous Underwater Vehicles

An AUV is an unoccupied, untethered, usually programmable underwater vehicle that is capable of roaming the ocean depths without pilot input. Although AUVs have been under development for several decades, they have progressed more slowly than ROVs, due mainly to technological challenges associated with their power sources and control. In the 1960s and 1970s, AUV development was funded primarily by the military for missions to search large areas under ice and in deep water over long time periods (NRC, 1996). It is only recently that applications have turned toward oceanographic science.

The AUV carries instruments that map the seafloor and measure a variety of physical and chemical ocean properties; they may transmit that information via a temporary connection to the launching station, surface at intervals to upload information via satellites, or store the information to be retrieved only when the AUV is physically recovered. By virtue of their relatively small size, limited capacity for scientific payloads, and autonomous nature, AUVs do not have the range of capabilities of HOVs and ROVs. They are, however, better suited for reconnoitering large areas of the ocean that could take years to cover by other means. AUVs are thus frequently used to identify prospective regions of interest that can be explored further with HOVs or ROVs.

SCOPE OF THIS REPORT

Over the last 20 years, a number of workshops have been held in which members of the U.S. deep submergence scientific community have discussed research priorities and assembled a "wish list" for needed new equipment (UNOLS, 1990, 1999). High on their list is a new, state-of-the-art HOV capable of descending to 6,000m or more. Deeper-diving AUVs and ROVs are also in demand. NSF/OCE is interested in assisting this very productive segment of the ocean sciences community and asked the

BOX 1-2
Statement of Task

This study will evaluate the future directions and facility requirements for deep submergence science and examine the range of potential applicable technologies that can support basic research in deep sea and seafloor areas.

Specifically, the Committee will:

1. Assess current and projected capabilities of occupied and unoccupied vehicles;
2. Make recommendations regarding the mix of vehicles needed to continue to carry out world-class deep submergence science; and
3. Discuss innovative design concepts and technological advances that should be incorporated into any new vehicles to support current and future research needs.

Recommendations will be made within the constraints (established by NSF) that total construction costs (including related costs such as modifications to support ships) shall not exceed the upper practical limit for the Division of Ocean Sciences' "mid-sized infrastructure" (roughly 10 percent of the division's annual budget or approximately $25M) and not require more than a modest increase beyond inflation in the operating budgets of the supporting agencies for deep submergence vehicles. If multiple types of vehicles are deemed necessary, suggestions will be made concerning the optimal phasing of implementation over several years. Cost estimates will be supplied by the National Deep Submergence Facility, other submersible owners and operators, or outside companies if needed.

National Academies to carry out an independent, objective assessment of the scientific and engineering needs and opportunities before making such a large infrastructure commitment (the formal statement of task can be seen in Box 1-2). In addition to the fiscal constraints specified in the statement of task, NSF/OCE indicated that capital investment will have to be made in the next two fiscal years if a new HOV is to be built during this decade (J. Yoder, National Science Foundation, Arlington, Va., written communication, 2003). Plans are currently under way to begin implementation of the research fleet upgrades recommended in a Federal Oceanographic Facilities Committee[2] report published in December 2001 and

[2]The Federal Oceanographic Facilities Committee is a federal interagency committee that operates as part of the National Ocean Partnership Program created by Congress in 1997 through enactment of Public Law 201-104.

entitled *Charting the Future for the National Academic Research Fleet: A Long Range Plan for Renewal* (Federal Oceanographic Facilities Committee, National Oceanographic Partnership Program, 2001). NSF/OCE is planning to cover the costs of constructing one Regional Class ship every two years beginning in FY 2006.

The purpose of this study is to provide NSF with recommendations for its consideration regarding activities to provide infrastructure support through NDSF or other means for basic research at depth in the oceans. As such, the discussions in this report are designed to inform this question and are not intended to provide an exhaustive account of all research-related activities carried out at depth or a complete account of all the potential assets that exist. The discussion of assets in this report is limited, therefore, to those that establish whether adequate DSVs exist within or without the NSDF. Furthermore any recommendations made in this report are above and beyond the needs of other large programs such as NSF's Ocean Observatories Initiative or activities falling within the realm of ocean exploration.

Approach and Information Needs

To evaluate future directions of deep submergence science in the United States, as well as the facility requirements and range of deep submergence technologies needed to conduct this science, a number of issues had to be considered. In an effort to evaluate options, the committee chose to systematically examine the question in terms of scientific need, technical requirements, necessary capabilities, and appropriate capacity. As the capabilities of both HOVs and unoccupied vehicles evolve, the demand for these platforms will also evolve (for example, if the enhancements recommended in Chapter 4 for NDSF's HOV are followed, the user base for that HOV will likely diversify and expand). The committee has declined to be drawn into a rhetorical debate about the proper mix of platforms 10 to 20 years in the future.

The DESCEND (DEveloping Submergence SCiencE for the Next Decade) report identified a number of laudable scientific goals but did not specify the unique role that deep submergence science would play or specifically what capabilities are needed to support targeted research to achieve these goals (UNOLS, 1999). There is a general and detailed list of desired capabilities, but these are not mapped to specific research initiatives. Input from the deep submergence community and a variety of oceanographic disciplines helped determine the science requirements, the geographic locations, depth ranges, and the current and future technologies needed for deep submergence science.

Understanding the limitations imposed on research carried out in the deep ocean requires knowledge of the mechanisms for awarding research funds and providing access to needed technology (e.g., as currently understood, obtaining funding and obtaining access to equipment involve separate processes). While *Alvin* and *Jason II* are currently oversubscribed, suggesting that demand exceeds availability, information on the pattern of use, criteria used to evaluate requests, rate of request denials, and relationship of scheduling access to resources maintained by NDSF was examined. Similarly, information on the funding proposal review processes used by NSF, likely level of funds to support research, and proposal success rates was also considered.

The technical capabilities for conducting deep submergence science have been examined extensively. Documents such as a 2001 article in the *Marine Technology Society Journal* on the development of undersea technologies (Rona, 2001), the 1999 DESCEND report (UNOLS, 1999), *Undersea Vehicles and National Needs* (NRC, 1996), The Global Abyss report (UNOLS, 1994), and *Submersible Science Study for the 1990s* (UNOLS, 1990) offer insights into the thinking of at least some subset of the user community. The present study expands on the perceived research needs and capabilities for the future. For example, the role of unique capabilities to enable high-priority science (i.e., how factors like human presence versus extended bottom time or range enable specific research efforts) is discussed and evaluated with respect to determining the depth capabilities and pattern of use for *Alvin* and other deep submersibles. Threshold depths and their corresponding geologic features (e.g., continental slope, abyssal plain, mid-ocean ridges, and deep trenches) are evaluated as an aid in determining the needs for deep ocean research platforms within the United States. The strengths and limitations of using HOVs, ROVs, and AUVs are discussed in detail to help specify the future needs for deep submergence assets as mapped to specific mission goals.

As the only NDSF HOV, *Alvin's* capabilities, strengths, and limitations have been evaluated (e.g., visibility, payload, bottom time, maneuverability) and recommendations made that consider deep submergence needs weighed against respective costs and benefits. Within the overall context of deep submergence science, use of *Alvin* is considered only as a component of an entire suite of DSV assets. In consideration of deep submergence needs and *Alvin's* important role, various options are provided that range from keeping *Alvin* as is, to improving it, to replacing it with a variety of different configurations. These replacement or modification options offer a range of improvements (e.g., improved bottom time, manipulator dexterity, data transmission, payload, and visibility) at various cost levels.

With adequate maintenance, *Alvin* could operate well into the foreseeable future. Although near-term replacement of *Alvin* may not be necessary, there is reason to believe that expanded capabilities are needed to support deep ocean research more fully. Potential construction costs for its replacement, as well as alternative designs and subsystem replacements for its upgrade, with consideration of annual maintenance and operating costs, are presented to inform a decision on whether to maintain, upgrade, or replace the *Alvin.*

Early in the study, the potential role of a full-ocean-depth HOV was raised. During subsequent discussions it was concluded that giving serious consideration to the potential construction and viability of an 11,000-m HOV was beyond the scope of the charge to the committee. The design and construction of such a vessel cannot be completed within the two-to three-year time frame NSF/OCE currently has to fund and initiate the construction of a possible HOV. For example, all of the existing deep-diving HOVs are designed around a sphere. To develop an HOV using proven designs and materials with similar occupant volume, while increasing the depth capability from 6,500 to 11,000m, would require doubling the weight of the sphere. For the new 6,500-m DSV, the titanium hull weight is approximately 11,000 pounds, or one-third the total HOV weight of 32,000 pounds. Doubling the sphere weight would nearly double the full-ocean-depth HOV weight and would place it well beyond the capacity of the support ship *Atlantis*. Modifying the *Atlantis* to handle a vehicle of such weight would place the total cost well beyond the NSF budget. Even if all of the components necessary to build an 11,000m HOV were available off the shelf, there is no certification test facility for such pressures. Given these limitations, it is simply not feasible for NSF to design and build a full-ocean-depth HOV for $25 million in two years. Given the short time available to provide NSF with advice, the committee focused on examining feasible options. Therefore, design approaches and the scientific value of a full-ocean-depth vehicle were not explored.

The potential utilization of non-U.S. facilities (e.g., HOVs from the Japanese Marine Science and Technology Center and the French Institute for Exploration of the Sea) was explored, but except for the Russian *Mirs*, which are available for lease, this option does not appear to be practical. The principal impediment is that the vehicles and their support vessels are typically fully scheduled with the needs of their home countries.

ORGANIZATION OF THE REPORT

The main focus of this report is to provide the evidence and arguments needed, as well as a range of options, to evaluate the greatest deep submergence vehicle capability for a set dollar amount. These issues are discussed at length and provide NSF with a list of possibilities for maintaining and improving NDSF assets.

Chapter 2 begins by defining deep submergence science. It documents the diverse nature of deep submergence science and describes the types of science conducted in the deep ocean as well as the geographic locations for current and proposed research. The hazards and difficulties of working at depth are outlined and the suite of available deep submergence platforms is introduced.

Chapter 3 further documents the strengths and limitations of various platform designs within the current suite of NDSF assets. An analysis of the number, suitability, and distribution of existing deep submergence assets (including non-NDSF vehicles) is made to support calls for expansion of available assets by improving current vehicles as well as adding other assets.

Chapter 4 explores the options for providing greater capabilities over a broader geographic range and articulates the justification for improving access to, and the utilization of, the nation's deep submergence assets. Additionally, options for upgrading individual components of ROVs and HOVs are presented as they relate to general mission goals. The chapter's focus is on improving the overall capability of the deep submergence fleet, including the standardization of tool sets and interfaces to be used on a broader range of deep-diving vehicles, HOVs and ROVs combined.

Chapter 5 brings together the individual findings to provide a coherent vision of how the agencies should support deep submergence science in the next 10-20 years.

Appendix A contains biographical sketches of members of the Committee on Future Needs in Deep Submergence Science. Although acronyms used in this report are redefined in each chapter, a complete list is provided in Appendix B. Appendix C contains a list of AUVs and their home institutions. Appendix D is a table of *Jason II* and the proposed HOV estimated subsystem weights and costs.

2

Understanding Deep Submergence Science

HIGHLIGHTS

This chapter

- Documents the diverse nature of deep submergence science (to establish the need for a mix of expertise and approaches)
- Documents the significance of this research for efforts to address some of the most compelling questions in science, not just ocean science (to demonstrate the soundness of supporting this effort despite the cost of working at depth)
- Describes the geographic extent over which this research must be done (to demonstrate the need to expand the ability to provide a suite of platforms over a wide expanse of the ocean)
- Describes the variable water depths at which this science must be done (to demonstrate the need for a suite of platforms capable of support work at a range of depths—not all on the seafloor)
- Presents a coherent and logical definition of deep submergence science (to support calls for consistent, equitable, and transparent mechanisms to provide access to scientific assets)

Research carried out at depth in the ocean over the last 40 years has provided dramatic and unique insights into some of the most compelling scientific questions ever posed. Understanding the nature of planetary processes and the fundamental constraints on the nature and existence of life has driven scientific inquiry to remote areas of Earth and the solar system. Yet when compared to the vast distances involved in space, the deep ocean lies essentially at "our back door." The remoteness and isolation of deep ocean environments makes this region of inner space a particularly fertile field for scientific inquiry. Fundamental contributions to the understanding of processes responsible for plate tectonics and ocean chemistry and physics, as well as the origins of life and mechanisms for speciation, have been made by scientists working at depth in the ocean. There is reason to believe that further discoveries can be expected if adequate access to these regions can be provided.

In the mid-1970s, during an investigation of geothermal plumes of water along a mid-ocean ridge spreading center of a tectonic plate, a completely unique community of life forms was discovered (Corliss et al., 1979). Hydrothermal vents, with their unique chemosynthetic communities and with their chemistry profoundly impacting ocean chemistry, are among the most important discoveries of the twentieth century. They have formed the basis for major research programs on the evolution of seawater and on the chemical, biological, evolutionary, and ecological relationships of vent organisms; they have revolutionized our understanding of processes controlling seawater chemistry and the origins and evolution of life on Earth; and they have stimulated new hypotheses regarding the possibility of life on other planets (Rothschild and Mancinelli, 2001).

The numerous scientific planning activities carried out by the Ocean Science Division of the National Science Foundation (NSF/OCE) and other organizations over the last decade have identified a number of reasons for investigating the depths of the ocean (J. Yoder, National Science Foundation, Arlington, Va., written communication, 2003). Public comments received during the course of this study from human occupied vehicle (HOV) and remotely occupied vehicle (ROV) users further identify areas of inquiry with significant scientific potential. (Examples of some of the comments received that pertain to deep submergence science are found in Box 2-1.)

Deep submergence science is a diverse field of study involving biological, chemical, geological, and physical oceanography, as well as marine archaeology. Habitats range from the vast midwater environments (Figure 2-1) to the ocean floor; from continents to the depths of the ocean basins; from plate boundaries at spreading ridges and ocean trenches to the remains of ancient and recent civilizations. The diverse nature of deep submergence science necessitates the use of a mix of expertise, approaches, platforms, and tools. There is an acute need for observations, sampling, and conducting of interactive and manipulative experiments in geographically diverse areas, from passive and convergent margins, to

Box 2-1
Deep Ocean Scientists on the Need for Deep Submergence Assets

First, let me emphasize the significance of structural geology and tectonics in the oceanic crust. Although these fields were born on land, deep submergence capabilities provide a means of bringing the techniques of field geology and structural analysis to the seafloor. If we are to understand these primary aspects of our planet, and use them to anticipate processes on other planets, it is essential that we push ahead with an aggressive program of deep-sea research supported with increasingly powerful research tools. — *Jeff Karson, Duke University*

Although the axis of most of the mid-ocean ridge is at depths of less than 4,500m, many of the transform faults fall within the 4,500 to 6500-m-depth range, particularly along the slow spreading ridges, which comprise more than 50 percent in length of the global ridge system. These areas expose sections of lower crust and upper mantle, hence allowing examination of the internal structure of oceanic crust. Such areas may expose both "fresh" and hydrothermally altered rocks (possibly including stockworks that underlie hydrothermal systems), and hence could provide valuable information about the hydrogeology and scale of subseafloor seawater convection. — *Susan Humphris, Woods Hole Oceanographic Institution*

The tectonic processes taking place in nonaccretionary forearcs, like the Mariana and Izu-Bonin convergent margins, can be studied at deep crustal-mantle exposures on the inner slopes of the trench and in deep faults in the outer forearc. There are several places where complete, intact sections (with throws of up to 5,000m) exist and expose the entire crust and upper mantle of the arc lithosphere within the active arc-forearc region. Such exposures provide new insights into the formation of arcs and processes that may lead to formation of continental lithosphere. — *Patty Fryer, University of Hawaii*

As the use of offshore resources presses to greater depths, the need for scientifically valid management strategies in the deep sea increases. Oil and gas development below 3,000m will soon be commonplace. Gas hydrate mining from depths of over 5,000 meters is under consideration for the future. Fisheries and ecosystems associated with these resources are an important component of any management or conservation regime, yet little scientific basis exists to develop sound management practices with respect to these resources. The regulatory agencies dealing with these pioneering issues, including the National Oceanic and Atmospheric Administration,

continued

Box 2-1 Continued

are developing strategies that in large part assume the deep sea is much like its more shallow counterpart. Unsound policies will be the long-term consequence. — *Barbara Moore, National Oceanic and Atmospheric Administration*

The overwhelming majority of the habitable living space in the ocean has not been adequately explored. There are still many surprises to be uncovered in terms of physical, chemical, and geological processes, but the potential for fundamental discovery in the biology of the oceans is particularly promising. Recently, we at Monterey Bay Aquarium Research Institute were quite surprised by the worldwide media attention devoted to Dr. George Matsumoto's and Dr. Dhugal Lindsay's paper describing a new species, genus, and subfamily of jellyfish—Big Red, or *Tiburonia granrojo.* Such surprising discoveries are happening all the time just by chance, and are likely to be even more commonplace if we become more systematic about exploring the ocean and documenting what we find. — *Marcia McNutt, Monterey Bay Aquarium Research Institute*

The Pacific Islands Fisheries Science Center has relied on both ROV and occupied submersibles to conduct research on deepwater assemblages of fish and corals. All of our work has been supported by the National Undersea Research Program through its Pacific node, the Hawaii Undersea Research Laboratory (HURL). The HURL fields both occupied submarines and an ROV, and we depend on both. — *Frank Parrish, National Oceanic and Atmospheric Administration*

ridge crests, ocean basins and the water column above, including as yet poorly studied ocean regions such as the Southern and Arctic Oceans. The ocean margins and ridges are ideal global laboratories to study dynamic interactions among physical, chemical, and biological processes. Submergence science provides a powerful way for conducting important research in the geosciences and biological sciences. The subsequent sections discuss only a few examples of the most compelling scientific challenges that call for access to the ocean depths.

NOTABLE AREAS FOR POTENTIAL CONTRIBUTION WITHIN THE GEOSCIENCES

Enhanced observation, sampling, and interactive experimental capabilities will lead to new findings and will thus open up new research per-

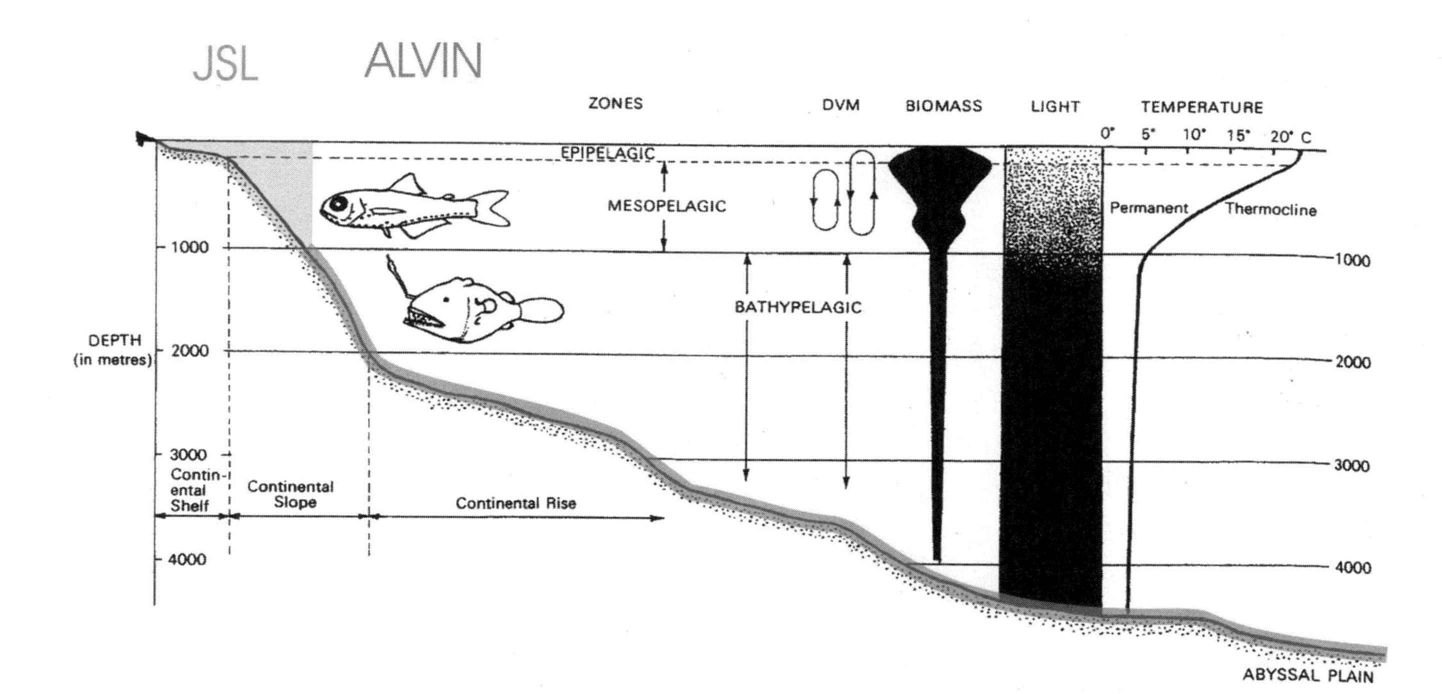

FIGURE 2-1 Ocean depth zones representing biological domains.
SOURCE: Widder, 2003

spectives and visions. For example, these findings will lead to a better understanding of interactions between the hydrological regimes in ocean margins and the consequences for earthquakes, slope stability, arc volcanism, ocean chemical balances, the global carbon cycle, clathrate hydrate formation and dissociation, ocean resources (e.g., hydrocarbon reservoirs associated with organic matter maturation in margins), and benthic chemosynthetic ecosystems sustained primarily by natural expulsion of nutrient-bearing fluids from overpressured sediments and rocks. Cold seeps, mud, and serpentine diapirs[1] and the associated chemosynthetic ecosystems were discovered with submersibles, and the great majority of what is known about this new frontier of science was obtained from the use of deep submergence vehicles.

In the late 1960s, scientific understanding of how Earth works was completely revolutionized by the newly accepted concept of plate tectonics and seafloor spreading. The surface of Earth consists of vast, rigid, lithospheric plates that are in relative motion. The driver of plate movement resides at depth, in the mantle. Large convection cells result in the slow motion of the mantle in the solid state. The mantle rises beneath mid-ocean ridges, melts, and erupts to form new ocean crust. The 60,000-km-long mid-ocean ridge system is therefore the location at which plates are generated. Newly created seafloor spreads at rates on the order of 1 to 20 cm per year. The seafloor generated at mid-ocean ridges covers two-thirds of the planet. Old seafloor is consumed at subduction zones, where two adjacent plates moving in opposite directions meet. The denser plate bends and plunges into the mantle, while the overriding plate is complexly deformed. At these convergent margins, the subduction of a plate causes differential partitioning of strain and generates earthquakes and volcanism. It is accompanied by dehydration reactions that release fluids, generating partial melting and thus, abundant magmatic activity. Eventually, the subducted plate is recycled into the mantle. Due to its constant renewing, the age of the oldest intact oceanic crust, located in the western Pacific, does not exceed 180 million years, and is therefore very young compared to the age of Earth.

This new concept was built on the visionary hypothesis of continental drift proposed by Alfred L. Wegener (1880-1930) at the beginning of the twentieth century to explain the match in shape and geological structure of continents across an ocean such as the Atlantic. In the plate tectonic scenario, continents do not drift but are anchored in the plate and entrained with it. However, because they are less dense than the ocean crust,

[1]"Diapir" refers to a general class of geologic structure that occurs when a lower-density layer of material is overlain by a higher-density layer of material. Salt domes are a classic example, as are shale diapirs and mud volcanoes.

they are not entrained in subduction. Consequently, the distribution of continents has changed during geological times. The initiation of plate rifting results in continental breakup: passive margins represent the transition between the continent and the newly formed ocean basin. The opening of a new basin eventually results in the closure of an older one and the collision of two continental masses. This process has occurred many times in the past and resulted in the building of successive mountain belts that have been subsequently eroded. At present, the closure of the Tethys ocean basin is responsible for the building of the Alps and the Himalayas.

A basic principle of plate tectonics is that plates are essentially rigid and do not deform. Therefore, most of the active areas are located at plate boundaries, where the plates are either constructed (mid-ocean ridges) or consumed (trenches associated with subduction). Another essential implication is that because these areas are located underwater, the understanding of processes governing plate tectonic requires exploration, measurement, and sampling of the seafloor. Among the various ways of investigating the seafloor, direct approach has proved to be critical. Mid-ocean ridges, passive and active plate margins, and hot spots are the key areas. Their investigation has benefited by, and still requires the use of, submersibles. All of the work accomplished at sea during the last 30 years has contributed to verifying the concept of plate tectonics. There are still a number of processes, however, that are not fully understood.

Mid-ocean Ridge Processes

With the widespread acceptance of plate tectonic theory, the value of direct observation of the seafloor became obvious to the whole scientific community. The next major step was the discovery of active hydrothermal vents at the end of the 1970s. "Black smokers" (Plate 4a) venting fluids at temperatures in the range of 350°C build up hydrothermal chimneys and sustain completely original chemosynthetic biological colonies that feed on chemical compounds dissolved in hydrothermal waters (Plate 4b). These vents result from seawater convection cells that are activated by magmatic heat, either from magma chambers located beneath the ridge or from cooling magmatic rocks. Mid-ocean ridges then became a major field of interest not only for earth scientists but also for fluid chemists and biologists (Plate 5). This new scientific excitement led to the creation of the Ridge InterDisciplinary Global Experiments (RIDGE) program by NSF/OCE to study the relationship between geological processes and the biology at mid-ocean ridge systems.

In the last 10 years, the focus of ridge studies has evolved from exploration and sampling toward in situ experimentation and long-term monitoring. This has been made possible through the development of instruments associated with submersibles—new types of samplers and sensors. It is clear that mid-ocean ridges function as systems; the aim now is to understand the linkages between what is happening in the mantle, the nature and distribution of RIDGE biospheres, and ultimately the composition of seawater. "From mantle to microbes" is the leading idea that drives the new RIDGE 2000 program. To answer this fundamental question, multidisciplinary experiments must be conducted at specific submerged locations. This has led the program to select a limited number of sites that represent the entire spectrum of ridge types: fast-; intermediate-; and slow-spreading ridges; hot-spot-influenced ridges; and backarc basins. In these sites, coordinated, multidisciplinary experiments by geologists, geophysicists, chemists, biologists, and oceanographers will require long-term monitoring to determine their temporal and spatial evolution. Some of these "integrated study sites" have already been selected by the community while others are still under discussion, but it is obvious that all will be at depths shallower than 4,500m. Because of improved research techniques these studies will require repeated cruises and more submersible time to conduct experiments and to deploy and recover instruments on the seafloor. Logistically, some of the selected sites (Juan de Fuca Ridge, East Pacific Rise) are located within the areas where *Alvin* normally works. The Lau basin, however, in the western Pacific, is very far away.

The RIDGE 2000 Science Plan (Ridge 2000, 2003) seeks to address the following questions:

- How is melt transport organized within the crust and mantle?
- How does hydrothermal circulation affect characteristics of the melt zone, the crustal structure, and ridge morphology?
- How does biological activity affect vent chemistry and hydrothermal circulation? What are the forces and linkages that determine the structure and extent of the hydrothermal biosphere?
- What is the nature and space-time extent of the biosphere from deep in the sub-seafloor to the overlying ocean?
- How and to what extent does the hydrothermal flux influence the physical, chemical, and biological characteristics of the overlying ocean?

These questions require submersibles to conduct detailed surveys and sampling programs, as well as long-term monitoring.

Convergent Margin Dynamics

Fluids contained in the subducting plate and overlying sediment and rocks are expelled at various levels under the influence of increasing pressure and temperature, often manifest at the seafloor as fluid seeps, and mud or serpentine volcanoes, associated with prolific chemosynthetic ecosystems (Kulm et al., 1986). Fluid is carried into subduction zones, both trapped in the pores of sediment and rocks and bound in hydrous minerals. The fate of the fluid varies from location to location (e.g. Oliver, 1986; Langseth and Moore, 1990; Peacock, 1990; Kastner et al., 1991; Moore and Vrolijk, 1992). At about half of all convergent margins, most of the sediment carried into the subduction zone is detached and added to the overriding plate to form an accretionary prism. Rates of fluid flow through these prisms are generally too low to significantly alter the thermal structures of subduction zones. It has become increasingly clear in many other ways, however, that the consequences of the presence and flow of fluids is profound. Where flow is sufficiently focused and fluxes of carbon and sulfur species are high, benthic biological communities are supported on energy derived chemosynthetically. In nonaccretionary subducting margins (e.g., the Mariana-Izu Bonin and Costa Rica subduction systems), the significant pore fluid volume in the underthrust sediments that is carried to a great depth (~15 km) potentially provides a large-volume source of fluid that is released primarily by compaction. Water bound in hydrous minerals in the sediment and altered oceanic basement is carried to even greater depths and is driven off by increasing temperatures and pressures. Decarbonation also occurs at greater depths. The relative importance of these various sources of fluids is beginning to be understood, but the quantities involved and the impact on ocean and mantle chemistries and microbiology are poorly constrained.

Early in the evolutionary stages of coarse-grained accretionary prisms, diffuse flow may be an important means of fluid expulsion (Kastner et al., 1993). In fine-grained, highly consolidated, or well-cemented sediments having low permeabilities, disequilibrium between production and drainage—hence overpressure or super-hydrostatic fluid pressures—evolve, causing fracture and fault zones to carry most of the flow (Carson et al., 1990). Under certain conditioning, hydrofracturing may result (Knipe and McCaig, 1994). Several chemical and isotopic tracers for tracking fluid flow, such as methane, chlorinity, and helium isotopes, are established. Fluids originating from compaction of sediments transported into the subduction system and from dehydration processes of the sediments and the subducting slab are expelled by tectonic consolidation and burial. Extreme pressures can create zero effective pressure (grain to grain contact) conditions and effectively fluidize sediments and rock, probably manifest in the formation of mud volcanoes. High fluid pressure may also reduce the effective strength of faults and earthquakes that cause fracturing, facilitating focused fluid flow. Submarine cold seepages, mud, and volcanoes (Fryer, 1992) are thus widespread in margin slopes and trenches, providing life-sustaining nutrients for large microbial communities within a broad range of temperatures, pressures, and diverse chemical environments, from methane-rich to sulfide-rich, and from seawater salinity to brines, most previously unknown. Some of the bacteria may become important for biomaterials.

Exploration is an extremely important component in both the RIDGE 2000 and the MARGIN programs. Only a small fraction of plate boundaries have been explored, and exciting new findings resulted each time a new segment was investigated. Although many similarities exist in the Atlantic and Pacific Ocean hydrothermal vent systems and associated benthic communities, they are rather distinct in some aspects. The same is true for margins. For example, the existence of serpentine and mud volcanoes and cold seeps was a surprising discovery. Their spatial distribution and geochemical significance are as yet unknown.

The following are but a few of the very intriguing and high-priority scientific questions dealing with convergent margins that can be addressed through deep submergence science (MARGINS, 1998, 2000a,b):

- What is the role of fluids in seismicity—in how earthquakes influence fluid flow?
- What is the mass exchange of fluid between the ocean and mantle, and what are the consequences of this flow for the balance of ocean chemistry, benthic biology, and mantle heterogeneity?
- How does the partitioning of fluid fluxes between the shallow outer forearc regions and depth determine where fluids contribute to melt production, arc volcanism, and mantle metamorphism?
- What is the influence of chemical fluxes on the mechanical properties and deformation of the overriding plate?
- What is the relative importance of transient versus steady-state hydrological processes? What is the role of organic carbon in the oceanic carbon cycle (margins account for more than 80 percent of the organic carbon in the ocean)?

Passive Margin Processes

Several important processes can be addressed using submersibles on passive margins. These include determining the mass flux of meteoric water versus seawater recycling in the ocean; the consequences of various flow regimes for ocean chemistry, benthic biology, and slope stability; the sedimentological and tectonic controls on hydrology; the influence of hydrology on accumulation and migration of hydrocarbons and gas hydrates; and the magnitude and spatial distribution of the driving forces on fluid flow. The hydrological regimes of passive rifted margins are largely unknown. Although fluxes, including those from meteoric systems on land, are predicted to be large as revealed by ^{226}Ra enrichment (Moore and Vrolijk, 1992), understanding of fluid flow regimes remains highly incomplete. Fluid flow in passive margins influences the migration and accumulation of hydrocarbons, and, thus, gas hydrate formation and accumulation, slope stability, and the morphology of the margins. Driving forces are derived from thermal pressure and salinity-density contrasts, as well as from topographic heads, but little is known about the magnitudes of the forces or the differences from location to location or with time. Rates of flow are only beginning to be quantified, in a few instances through the use of tracers from meteoric water (e.g., Cable et al., 1996). The lateral and depth extent of passive margin fluid flow systems is not well understood. Major advances in understanding these processes, and the environmental consequences will depend on accessibility to existing and new platforms for deep submergence science. Direct measurements of flow rates through sediments, at focused and nonfocused flow sites (using flowmeters deployed by submersibles, for example), can provide essential information on several of the processes mentioned above. Sampling bottom waters in areas of complicated bathymetry represents a technical challenge for conventional wireline water sampling equipment. ROV-mounted isotopic systems would be ideal for overcoming this challenge.

The potential for a large-scale failure of the shelf upper slope, off Virginia and North Carolina, where landslide scars exist, is an example of the acute need for access to innovative submergence science in an environment and geographic region previously neglected. Enigmatic asymmetric cross-cutting "crack"-like features up to 50m deep, arranged in an echelon fashion along a 40-km section of the shelf edge have been documented with high-resolution side-scan seafloor imaging chirp subbottom profiling. The cracks appear to result from massive expulsion of gas through the seafloor, creating permeable pathways for updip-upslope gas migration. Gas accumulations in marine sediments have profound global environmental, geotechnical, hazard, and resource significance. Using submersibles to study these features along the shelf edge of the mid-Atlantic margin would provide data on new, previously unknown, methane venting processes and sites in the marine environment.

Ocean Geochemistry

The mass flux of fluid into the ocean from cold seeps and ridge flanks, and from elevated temperature seeps and hot vents at ocean ridges and hot spots, profoundly influences seawater chemistry. A volume equivalent to the entire ocean cycles through the ridge crest and off-axis venting systems in 1 million to 3 million years and through convergent margins in a few hundred million years.

Fluid expulsion through the ocean sediments and basement is driven by temperature, pressure, and density gradients. These fluids support rich benthic biological communities on energy derived chemosynthetically, mostly from sulfur and carbon compounds. Life on Earth may have originated at such seeps and/or vents. Global fluid fluxes and their impact on ocean geochemical budgets are unknown. Instrumenting and sampling focused fluid flow sites for short- and long-term monitoring of chemistry and fluxes, and observing the effects of flux fluctuations on the benthic and benthopelagic communities, require submergence facilities (i.e., vehicles, special sensors, specialized instruments). The seeps and hydrothermal vents span a wide depth range, from a few hundred meters at margin slopes, to ~2,500m at ridges and >6,000m at trenches.

The fluids emanating from such focused sites into the ocean propagate laterally and vertically. Observing their interactions with seawater, in particular the rates and processes leading to mineralization and scavenging of trace elements (e.g., phosphorus by iron oxyhydroxides), and their effect on the near-bottom and midwater biological communities require submersibles with special capabilities.

Gas Hydrate Occurrence and Formation

Clathrate hydrates are crystalline compounds in which an expanded ice-like lattice forms cages that contain gas molecules. The hydrates are stable only when gas molecules occupy the cages at moderate to high pressures and low temperatures. Their stability also depends on the composition of the gas molecule and the pore fluid chemistry (Handa, 1990; Sloan, 1990; Dickens and Quinby-Hunt, 1997). Methane hydrate is the most common natural gas hydrate on Earth (Sloan, 1990; Kvenvolden, 1995). Pressure and temperature constraints and the availability of methane restrict methane hydrates to two main environments on Earth: (1) in the modern ocean temperature regime in water depths exceeding 500 m (shallower beneath colder Arctic seas), and (2) on land beneath high-latitude permafrost. Even conservative estimates, within an order of magnitude, converge on the value of ~10,000 gigatons (21×10^{15} m^3) of methane (Kvenvolden, 1988), approximately 3,000 times the amount of methane in the atmosphere. The mass of carbon (C) in methane hydrate is about twice the amount of all fossil fuels on Earth, and thus comprises a major pool linked to the oceanic carbon budget.

Methane hydrate is considered important for a variety of reasons. First, its dissociation may be caused by natural tectonic, oceanic, and/or climate processes in the sediments that can lead to large-scale instability by creating overpressures that can trigger submarine landslides, thus controlling natural seafloor mass movements and generating tsunamis (Kayen and Lee, 1991; Booth et al., 1994). Understanding the dynamics of gas hydrates is important for ocean chemistry and biology. Catastrophic landslides caused by bottom-water temperature and/or pressure fluctuations could abruptly release much methane from dissociating hydrates to the ocean, impacting the dissolved oxygen content in the water column.

Second, when released to the atmosphere, methane is a powerful greenhouse gas that may have had impact on past climates. Along with CO_2 and water vapor, methane is one of the most effective greenhouse gases. Because of the potential for hydrate decomposition as a consequence of a greenhouse gas-induced global warming, a small change in stability regime, such as bottom water temperature variations, could result in large and potentially rapid climatic feedback effects (e.g., Raynaud et al., 1993; Behl and Kennett, 1996). If only a small fraction of the methane hydrate in the world occurs in extractable concentrations, it might be a significant energy resource for hydrocarbon fuel. Although it will continue to increase the global greenhouse effect in the atmosphere, as a fuel it would be significantly less polluting than other fossil fuels.

Because methane hydrate is unstable at typical surface temperature and pressure and is highly susceptible to dissociation when perturbed, studies of the dynamics of its nucleation, formation, and dissociation in response to environmental perturbations are very challenging. They require careful new observations, sampling, and recovery at in situ conditions. Novel seafloor and water column measurements and experiments, such as in situ interactive and perturbation experiments, as well as development and testing of new measurement tools, are essential and require the use of submersibles.

Studies of gas hydrates in the natural environment involve the skills of a diverse community of scientists and the use of a wide range of existing and newly developing techniques. Some examples of these techniques are high-resolution geophysical seafloor observations and mapping systems; uncontaminated chemical and biological sampling under natural conditions and in situ biogeochemical analyses; and development and testing of new in situ logging for physical and chemical parameters, such as nuclear magnetic resonance (NMR) measurements. Careful integration of well-designed, long-term field studies with an array of submersible vehicles, particularly ROVs closely linked with autonomous underwater vehicles (AUVs), is needed in order to refine quantitative estimates of gas and hydrates in the continental margins, to test various hypotheses for the formation of hydrates, and to understand the circumstances and environmental consequences of their dissociation.

NOTABLE AREAS FOR POTENTIAL CONTRIBUTION WITHIN THE BIOLOGICAL SCIENCES

Discoveries made using submersible vehicles have radically altered our understanding of life on Earth, from its origins and present diversity to its evolutionary processes and its future. Underlying most of the biological research in the deep sea are questions of ecological structure and dynamics, whether the fauna occurs on the abyssal plain, in the oceanic water column, at a cold seep, or at a hydrothermal vent. The following important questions remain to be studied and answered:

- How are these communities structured and how are they organized into functional ecological groups?
- How have the constituent species adapted to conditions in these habitats that appear to us to be so extreme?
- How can we use what we learn about such adaptations for human benefit?

Evolution and Ecology of the Deep-Sea Benthos

Much of the current research is focused on understanding the evolutionary and ecological processes that determine the spatial and temporal structure of deep-sea communities. For benthic communities, this includes studies of larval dispersal, patterns of diversity, colonization, and speciation (Plate 6a,b). The intellectual drivers for studies of evolution at hydrothermal vents and cold seeps—and, by extension, other deep-sea habitats—can be summarized by four questions:

1. What are the forces that direct the evolution of chemosynthetic communities and of the deep-sea fauna in general?
2. How do chemosynthetic fauna disperse between ephemeral seafloor localities that are isolated along or between ridge axes?
3. Do topographic features such as transform faults disrupt gene flow along a ridge system?
4. Do cold seeps, wood, carrion, and whale falls provide stepping stones between isolated vent habitats? Are they responsible for the observed levels of diversity?

These questions and attendant ecological studies of the factors that structure the diversity of deep-sea communities require in situ observations and experimental manipulations. Capabilities necessary to support this work include the following:

- High-resolution quantitative imaging and mosaic mapping
- Manipulative functions for placement of experimental packages and arrays
- Discrete, high precision measurements of fluid chemistry at a range of temperatures
- In situ, real-time characterization of habitats in terms of temperature, chemistry, geology, and flow
- Sampling tools to provide faunal collections for genetics and for productivity estimates.

Increased exploration reveals that biodiversity in the deep sea is far greater than could have been imagined even a decade ago. Understanding the broad-scale patterns and processes of biodiversity, on the seafloor and in the water column, is one of the most challenging areas of modern deep-sea biology.

Carbon Dynamics Linking Midwater and Benthic Communities

A fundamental puzzle of deep-sea biology concerns the transfer of energy from the primary producers at the top of the water column to the animals that inhabit the deep seafloor below, over an average distance of 4,000m. Measurements of this transfer are traditionally based on sediment traps, which collect the slowly sinking, small particles of detritus that are believed to be the principal source of food for the deep. Recently, measurements made by deep submergence vehicles of the metabolic energy being utilized by deep benthic communities revealed a significant discrepancy. Substantially more energy is being utilized than can be accounted for by the traditional means of measuring its arrival (Smith and Kaufmann, 1999).

The challenge of this situation is to accurately trace the routes and rates of energy transfer through the water column and its complex, resident midwater communities and accurately measure the input to the benthic community. Preliminary evidence, also acquired by deep submergence vehicles, suggests that the spectrum of detritus that reaches the benthos includes large, rapidly sinking particles that are not collected by sediment traps. The in situ perspective provided by undersea vehicles is helping to resolve this global-scale issue, just as these vehicles initially revealed the problem (Druffel and Robison, 1999).

At the interface between the base of the oceanic water column and the deep-sea floor is a region of heightened biomass and resuspended particulate matter known as the benthic boundary layer. Within this "benthopelagic" transition zone live a variety of both active swimmers and passive drifters. In addition, there is a potentially enormous biomass from microbial populations of the underlying abyssal and hadal sediment column and associated basement rock. Because this community is vertically narrow, it is difficult to tow nets accurately within its boundaries. Soft mesh nets suitable for capturing delicate pelagic animals seldom survive contact with the bottom, and trawls rugged enough to survive contact typically destroy the soft-bodied fauna they catch. What is required to accurately study this region are vehicles that can measure and control their height above the seafloor, while observing, recording, and sampling the resident fauna.

Ecological Structure of Midwater Communities

Between the sunlit upper layers of the open ocean and the dark floor of the deep sea is the largest living space on Earth. Within this immense midwater habitat are the planet's largest animal communities—largest in terms of distribution area, numbers of individuals, and biomass. It is the least explored of all the Earth's major habitats. Midwater animals are adapted to a fluid, three-dimensional world without solid boundaries (Plate 7a,b). Conducting research in this context presents unique challenges to scientists and to the tools they use.

The ecological structure of midwater communities is rooted in the phytoplankton that occupy just the upper 2 to 5 percent of the water column. The sunlight that powers photosynthesis near the surface also serves to illuminate the midwater habitat, at least during daylight hours. The eyes of visually cued predators are surprisingly effective even at dimly lit depths. As a result, a large number of midwater animals follow changes in the light field to stay at light levels optimal to their predator avoidance and prey capture strategies. Although these daily vertical migrations cover at most several hundred meters per individual, cumulatively they are the largest mass migrations on Earth and are a profoundly important dynamic aspect of oceanic ecology. Yet while we have been aware of these migrations since the advent of SONAR (sound navigation and ranging), we know surprisingly little about them. What is needed is the ability to track and observe the migrators throughout their daily cycles, to monitor their activity, and to examine aggregations and species interactions. The means to accomplish these goals are undersea vehicles that can travel through the water column as freely as the animals themselves, with a degree of stealth that will not disrupt the natural patterns.

Another common adaptation to the reduced midwater light regime is transparency, a highly effective predator avoidance strategy. As much as a third of all macroscopic animals in the upper 1,000 meters are essentially invisible. Imaging systems with side-lighting, back-lighting, or polarization capabilities, in combination with human eyes and high-resolution cameras, are required for observations and enumeration.

Communication in the Deep Sea

Biologically produced light, or bioluminescence, appears to be one of the most common form of communication in the oceanic water column and, by extension, on Earth (Widder, 1997). Midwater animals use bioluminescence in a variety of ways. While the diversity of luminous organisms has been revealed by conventional sampling methods, the dynamics of bioluminescence are still unknown. The most promising method for understanding the full scope of bioluminescent communication in the deep-sea will be through in situ work with undersea vehicles through direct observation of the behavior of individual organisms and the interactions of multiple organisms that can be compared to bioluminescent patterns.

Communication via pheromones and other chemical cues is also thought to be widespread. Such excreted chemicals travel rapidly along fine-scale horizontal structures in the water column. Mapping of such vertically stratified fine-scale structures along a horizontal plane can be done only with nontethered platforms incorporating a real-time feedback mechanism. Autonomous vehicles and HOVs are the ideal platforms for such studies.

Gelatinous Animals and the Pelagic Food Web

One of the most important discoveries by midwater researchers in recent years has been that gelatinous animals form a dominant ecological component of midwater communities worldwide (Plate 8a,b,c,d,e,f). The segments of the pelagic food web occupied by jellyfish (cnidarians, ctenophores, etc.) includes at least three trophic levels, from primary consumers to top-level carnivores. Because they are perfectly adapted to their fluid habitat, these animals are soft bodied and frequently very fragile. As a result, they were seriously underestimated by conventional sampling methods such as trawl nets. It was not until midwater-capable undersea vehicles became available that the extent and importance of this "Jelly Web" was realized. This part of the food web is a partially closed system because so many gelatinous animals feed on other jellies, thus sequestering a substantial fraction of pelagic biomass in their bodies. Even less well known is how the organic material consumed by this portion of the overall web is cycled back into the rest of the deep-sea community. The copious amounts of mucus produced by some gelatinous animals and their ability to capture small particulates hint at a major role in marine snow production; cnidarians are also known to take up dissolved organic matter. The significance of this aspect of midwater ecology requires further exploration and discovery. Just as undersea vehicles were critical to opening these lines of research, they will be vital to the measurement and process studies to follow.

Speciation and Biodiversity

The evolutionary, geological, and historical factors driving speciation in the pelagic environment are radically different from those operating in terrestrial and seafloor communities due to the fluid nature of ocean currents and consequent mixing effects. On a geological time scale the isolation of communities from each other can be ephemeral, and understanding how speciation occurs in the open ocean, particularly in the more stable deep sea midwater environment, will undoubtedly force us to rethink paradigms based on terrestrial systems. Factors contributing to the extremely high diversity seen in pelagic communities, concerning the number of species coexisting at any given survey point, can be elucidated only when the entire community is sampled. For the gelatinous component of this community, this is possible only using undersea vehicles. The sampling of gelatinous organisms using submersible-mounted equipment allows fragile animals to be collected in pristine condition, which in turn allows accurate taxonomic data to be collected for these animals. It is only with accurate taxonomic data that cryptic species, subspecies, and life stages of some organisms can be identified. Without these data we cannot have an accurate description of the biodiversity in the midwater environment.

Just as *Alvin* revolutionized deep-sea benthic biology and geology by providing an in situ perspective, submersibles better suited to operate in the midwater are enabling scientists to radically change the way we see and understand that vast ecosystem. These HOVs and ROVs provide access to the midwater habitat to enable direct observation and intervention for research. This new viewpoint lets us see midwater animals in the context of their environment, instead of being hauled to the surface in the bottom of a net. It allows studies of the dynamic aspects of biology, including behavior, in situ physiological measurements, and the interactions between species—none of which were possible before.

OCEAN EXPLORATION

An important element in the use of deep submergence vehicles is ocean exploration.

A recently released National Research Council report, recognized the necessity of an ocean exploration program to identify and describe the ocean's resources. Such a program would "provide opportunities for investigating new regions and that draws on research from a variety of disciplines, would speed discovery and application of new information," including climate change predictions, beneficial new products, informed policy choices, and enhanced ocean stewardship (NRC, 2003b, p. 3).

The report pointed out that in the United States ocean scientists "rely on relatively few, large, carefully managed assets—ships, submersibles, and laboratory facilities." Because the assets available to ocean science are already stretched thin, any new ocean exploration program that would enhance current efforts "will require substantial assets. New . . . assets would increase the effectiveness of the program, while minimizing interference with the current research endeavors" (NRC, 2003b, p. 14).

As required by its statement of task, the committee on Exploration of the Seas developed budget scenarios for supporting an exploration program. Three levels of support were described, and the highest level of support recommended including "a ship, three HOVs, five ROVs, and ten AUVs " (NRC, 2003b, p. 140). In addition, a separate flagship capable of supporting simultaneous ROV and HOV activities was recommended to "maximize program capabilities [and] ensure access and scheduling flexibility" (NRC, 2003b, p. 143).

It is important to note that any assets required by an enhanced ocean exploration program are above and beyond those discussed and recommended in this report.

FUTURE NEEDS

Advances in all of the areas discussed above will require highly focused, high-resolution studies that integrate results from various disciplines. The selection of representative sites must be based on scientific objectives and not dictated by the need to do all research in restricted geographic regions. Great effort must be made to find examples of processes that are simple as well as representative so that results are as unequivocal as possible and have utility in a global context.

To bear fruit, the majority of the research described above requires only greater access to deep submergence platforms. In some instances, especially midwater work or work at greater depth, research will require more advanced platforms with greater capabilities. Development of an adequate pool of assets necessary for continued deep submergence research will require long-term commitments. A well-equipped oceanographic fleet, including HOVs, ROVs, and AUVs, is well beyond the development phase, and its worth to a broad base of marine scientists has been well demonstrated. In many cases, the development and testing of highly specialized and novel instrumentation will be required. In others, experiments may require several years to complete. Clearly some means of achieving an appropriate balance must be found to provide continuity for longer-term projects, experiments, and responsiveness to promote innovation and access to support monitoring efforts.

3

Overview of Existing and Planned Assets

HIGHLIGHTS

This chapter

- Presents the argument that the current suite of assets will limit efforts to achieve the scientific potential discussed in Chapter 2 (to support calls for changes in the number, nature, and accessibility of deep submergence assets)
- Documents the strengths and weaknesses of various classes of platforms currently available (to support recommendation of a mix of assets and need for improvements of subsystems)
- Describes the number, suitability, and distribution of existing assets (to support calls for expansion of available assets by improving access to non-NDSF assets and adding new assets)

Research efforts to gather data from the undersea world have their origins in surface ship techniques for obtaining samples from beneath the ocean's surface. It became rapidly obvious, however, that in situ observation and sampling was crucial to understanding processes occurring at depth. This led the scientific community to use undersea vehicles that was developed for the military, for industry, or specifically for academic research. Three categories of undersea vehicles are now commonly used:

1. HOVs (human-occupied vehicles) are autonomous vehicles driven

by a pilot, and generally with a crew of two (two scientists or one scientist and a copilot). They are equipped with sampling devices and multiple sensors. Although often augmented by video systems, they also allow direct observation with the human eye through viewports or a transparent acrylic sphere. The duration of a dive is limited by battery life, human endurance, and safety protocols (e.g., operation only during daylight hours), and typically does not exceed 10-12 hours, including descent and ascent. The exceptions are the Russian *Mir* submersibles operating on a 100-kW battery that can provide dive times in excess of 14 hours. Although more than 200 HOVs have been built worldwide since the late 1950s (NRC, 1996), only about 16 are used for scientific investigations worldwide. Four countries presently operate HOVs with a capability to dive more than 1,000m for scientific purposes: United States, Japan, France and Russia (Table 3-1).

2. ROVs (remotely operated vehicles) are controlled by a pilot outside the vehicle rather than within it. Generally the pilot works from a surface ship to which the ROV is tethered by an "umbilical cord." This umbilical cable provides electrical power and control commands to the ROV and transmits sensor and feedback data to the pilot. ROVs are often operated from a sophisticated control room where monitors display the images of the seafloor or water column in real time. The advantage of the control room is that it allows a number of scientists or engineers to discuss the data in real time and make collective decisions about the operations. Because of the umbilical cable, there is no energy limitation and the vehicle potentially can remain below the surface for days. The umbilical cord, however, represents a constraint on operations because the range of the vehicle with respect to the ship cannot exceed a few hundred meters (some fraction of the length of the cable). The ship must therefore anticipate the movements of the vehicle, which requires a ship equipped with a dynamic positioning system. Moreover, the presence of a tether limits maneuverability and can introduce entanglement problems in rugged terrain such as hydrothermal fields with high chimney structures (up to several tens of meters). There are a large variety of ROVs with various sizes and depth capabilities. ROVs are used routinely for offshore oil and gas operations for the support of subsea cable laying, retrieval, and repair. ROVs also have been developed or purchased for scientific purposes in many countries.

3. AUVs (autonomous underwater vehicles) are unoccupied submersibles without tethers; all power is supplied by onboard energy systems. They are generally fitted for specific tasks, and their mission is programmed into the AUV before launch. Remote control communication, although possible with acoustic transmissions, is not practical in the absence of a cable. Some AUVs can operate as hybrid ROV-AUV systems

TABLE 3-1 Major Existing HOVs

HOV	Operator	Maximum Operating Depth (m)
Shinkai 6500	Japan Marine Science and Technology Center (JAMSTEC)	6,500
Mir I and *II*	P.P. Shirshov Institute of Oceanology, Russian Academy of Sciences	6,000
Nautile	French Research Institute for Exploitation of the Sea (IFREMER)	6,000
Alvin	National Deep Submergence Facility (NDSF), Woods Hole Oceanographic Institution (WHOI), United States	4,500
Cyana[a]	IFREMER, France	3,000
Shinkai 2000[a]	JAMSTEC	2,000
Pisces IV	Hawaii Undersea Research Laboratory (HURL), United States	2,170
Pisces V	HURL, United States	2,090
Johnson Sea-Link I and *II*	Harbor Branch Oceanographic Institution (HBOI), United States	1,000
Deep Rover	Nuytco Research Ltd., Canada	900
Remora 2000[a]	Comex, France	610
JAGO	Max Planck Institute, Germany	400
DeepWorker 2000	Nuytco Research Ltd., Canada	600
Delta	Delta Oceanographics, United States	370
Clelia	HBOI, United States	300

[a]Not currently in use.

(Japan Marine Science and Technology Center's AUV *Urashima*) through use of a single microfiber-optic cable. For nonhybrid AUV systems, the data are recorded and recovered with the vehicle, while a subset can be sent to the support ship via acoustic transmission. A large variety of AUVs are available, and they are used for military, industrial, or scientific purposes.

ROVs with an extremely long cable (e.g., ROV *Kaiko*) typically require a dedicated mother ship with a dynamic positioning system and a suitable handling system to launch and recover these vehicles. According to the National Deep Submergence Facility (NDSF), the daily rates of operation for the *Alvin* and *Jason II* are similar, and both HOVs and ROVs require a technical team of three to six individuals (this is similar to running a modern working class ROV, which requires a minimum of three people for 12-hour operations). If there are special tools, a fourth person may be needed. Twenty-four-hour operations require six-person crews. Rental rates for commercial ROVs vary considerably depending on the type of system, how it will be used, and the skill level and experience of the crew. Each mission is different, and therefore a direct comparison of costs to run commercial and NDSF assets is not straightforward. For a given cruise duration, ROVs actually spend more time on the seafloor or water column than HOVs because they do not have to return to the ship every day. The operation of an AUV typically requires three to four people.

HUMAN-OCCUPIED VEHICLES

Although many HOV submersibles are capable of providing greater scientific access to the deep ocean, there is some question about whether they are being used to their fullest extent. The following are brief descriptions of some of the most frequently used and best-known deep submergence vehicles. Although some of these vehicles are currently available for users outside their home institutions, a comparison of safety standards with those of U.S. NDSF vehicles was outside the scope of this study. Scientists proposing to use non-NDSF deep submergence assets are encouraged to review those safety issues.

Alvin

The *Alvin* is the only NDSF HOV and is operated by Woods Hole Oceanographic Institution (WHOI). It is owned by the U.S. Navy and is usually contracted through funding by the National Science Foundation (NSF). *Alvin* has one pilot with room for two scientific observers and is rated to a maximum operating depth of 4,500m (see Figure 2-1). *Alvin* is fitted with two manipulator arms, and four externally mounted video cameras; it has both quartz iodide and metal halide lights, and has a front sample basket that can hold 454 kg (1,000 pounds in air) of specimens and gear.

At present, *Alvin* is scheduled at 100 percent of her available time, which is limited by a maintenance schedule. As shown in Figure 3-1, the number of dives *Alvin* conducts each year varies somewhat, but it rou-

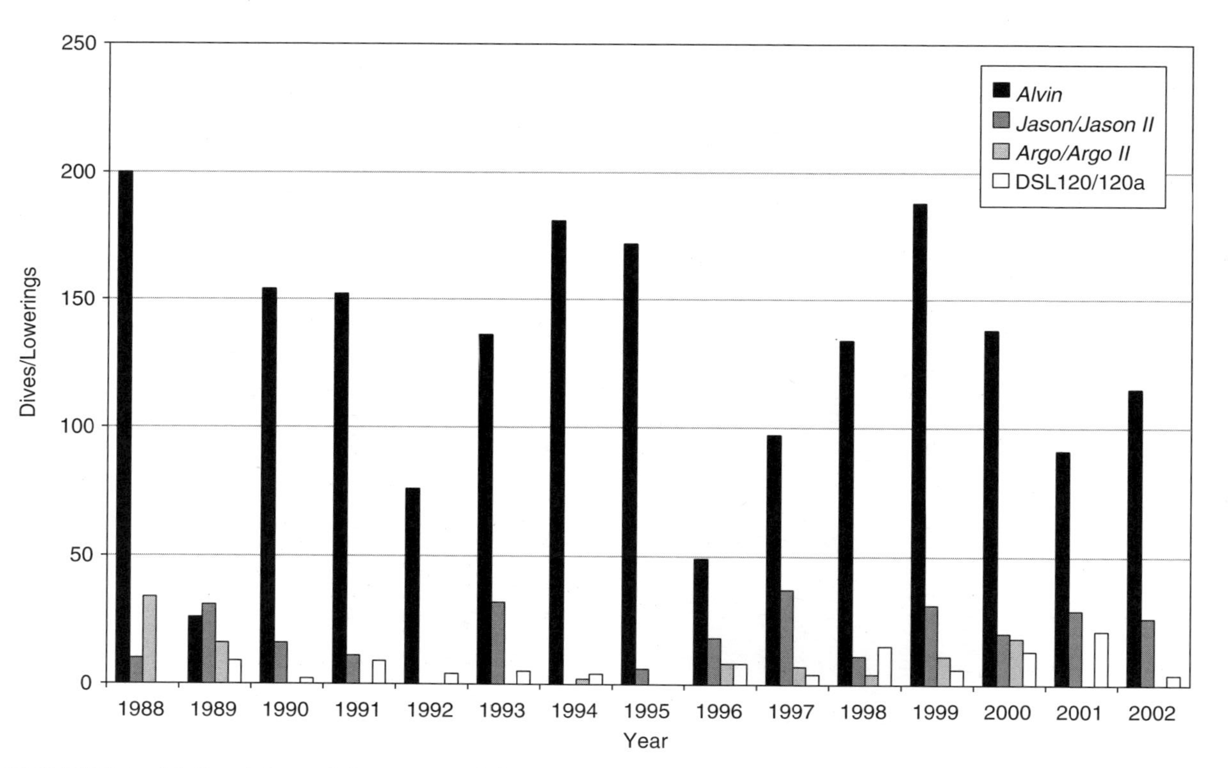

FIGURE 3-1 National Deep Submergence Science Facility vehicle dives or lowerings per year since 1988.
SOURCE: Data from the National Deep Submergence Facility, Woods Hole Oceanographic Institution.

tinely averages between 100 and 175 dives each year. Dive schedules are dependent on overhauls and maintenance, although some potential time is lost by transiting to various dive locations. Arrangements for use of *Alvin* are made through the University-National Oceanographic Laboratory System (UNOLS) NDSF.

Shinkai 2000

The *Shinkai 2000* is operated by the Japan Marine Science and Technology Center (JAMSTEC) for use by JAMSTEC scientists. At present there is no funding by JAMSTEC for this vehicle, which is currently in storage but is considered fully capable and ready to dive on very short notice. Use of the *Shinkai 2000* submersible by JAMSTEC or outside parties would require funding a full year of operations. This has resulted in little or no interest by either JAMSTEC or outside parties in making use of this vehicle.

Shinkai 6500

With the *Shinkai 2000* in extended dock, JAMSTEC now operates only one occupied submersible, the *Shinkai 6500*. This vehicle has a titanium hull and a maximum operating depth of 6,500m, making her capable of visiting approximately 98 percent of the world's ocean volume. She was built in 1989, requires a pilot and a copilot, and has room for a single scientific observer. Two manipulator arms and one forward-looking and two forward and down looking viewports provide the pilots and observer with partially overlapping fields of view. The seafloor off the Japanese coast is a collection of subducting lithospheric plates, an 8,000m trench, and an abyssal ocean bed with accumulated sediments, all of which provide a multitude of scientific research possibilities close to *Shinkai 6500's* home port.

Specifications for the submersible include a titanium pressure hull, a weight of 25.8 tons in air, a maximum operating depth of 6,500m, a maximum speed of 2.5 knots, and two three-chip CCD (charge-crupled device) video cameras, one fixed to the front of the vehicle and the second on a pan-tilt unit between the pilot's and observer's viewing windows. The *Shinkai 6500* is outfitted with seven lights: four 500-W halogen lamps, two 1,000-W halogen lamps, and one 250-W thallium lamp. Video footage is recorded onto digital video (DV)-Cam videotapes. There are two manipulator arms (7 degrees of freedom) and two large movable sample baskets. Payload weight is 200 kg in air. Regular dives last up to 9 hours, with life support available for 129 hours.

Currently the *Shinkai 6500* dives approximately 60 to 90 days per year due to an extremely rigorous maintenance schedule, which is necessary given the extreme depths to which she dives. Although arrangements between JAMSTEC and non-JAMSTEC co-principal investigators (PIs) do occur, they are not the norm. For these reasons, use of the *Shinkai's* is not considered a feasible option when NDSF assets cannot be used.

Johnson Sea-Links I and *II*

The use *Johnson Sea-Links I* and *II* (JSLs) are owned and operated by Harbor Branch Oceanographic Institution (HBOI). The JSLs (see Plate 1b and Figure 2-1) are classed and certified to a maximum operating depth of 914.4 m by the American Bureau of Shipping (ABS). The forward 5-inch-thick acrylic sphere accommodates the pilot and a scientist at 1 atmosphere (atm) and allows panoramic visibility. A second crew member and another scientist occupy the aft observation chamber where a video monitor and side viewports provide forward and side observations. The evolution of specialized equipment such as manipulator arms, suction devices, and rotary plankton samplers has made it possible to accomplish almost any work from within the submersibles that could once be done only by divers in shallower depths. The JSL submersibles are further outfitted with active SONAR (sound navigation and imaging), laser-aimed, still and broadcast-quality video cameras and HBOI-developed xenon arc lights. JSL I and II have been in operation since 1971 and 1975, and have completed 4,545 and 3,400 dives, respectively. Typical applications for the JSLs include benthic and midwater observations; photo or video documentation and collection of organisms; punch and box coring; search and recovery; bottom surveys; photogrammetric surveys; archaeological site documentation and recovery; and environmental impact studies. The combination of panoramic visibility, a variable ballast system, and a variety of collection tools make the JSLs especially well-suited for midwater studies.

Although HBOI researchers and collaborators from other institutions are the primary users of the JSLs, the vehicles are available for other users. In fact, a significant portion of the research conducted by the JSLs has been funded by agencies such as the National Science Foundation (NSF), National Oceanic and Atmospheric Administration (NOAA), and Office of Naval Research (ONR). NOAA and ONR provide support for JSL operations; although NSF supports the research and ship operations conducted using the JSLs, it supports operation only of the NDSF HOV (*Alvin*). Use of the JSLs is therefore a prospective option when NDSF assets are not available or appropriate.

Nautile

The French Research Institute for Exploitation of the Sea (IFREMER) submersible *Cyana* has been permanently retired. The sub was built mostly for observation and has a depth capability of 3,000m. The suite of scientific missions has shifted from observation to sampling and experiments, and *Cyana* is not well suited for that type of activity. An improvement over *Cyana* was *Nautile*. Like *Cyana*, *Nautile* allowed a single observer and required a pilot and copilot. *Nautile* has a depth capability of 6,000m, two manipulators, a large sampling basket, and an equipment payload of 200 kg. The two major portholes (for the pilot and the scientist) are located in front of the vehicle and their field of view overlap, which makes sampling operations easier. *Nautile* recently underwent a refit that resulted in major improvements of the navigation and camera systems. It is occasionally used for nonscientific purposes such as filming historical shipwrecks, salvage work on the *Titanic*, and work on the wreck of the oil tanker *Prestige*.

The *Nautile* was constructed and is operated by IFREMER for its scientists through its operational organization GENAVIR. Proposals for cruises are evaluated at the national level by independent committees for the various scientific disciplines. IFREMER then constructs a schedule that accommodates the ranking of the proposals, the logistics of the ships, and the budget of the fleet. This schedule is approved at the national level by an independent committee. There is presently a move to evolve from an annual schedule to a multiannual schedule that would facilitate the logistics of the ships. Due to budget limitations, neither of the HOVs is presently used to its full potential. *Nautile* dives only approximately six months per year. To increase the budget of the fleet, *Nautile* has been leased occasionally for commercial purposes. IFREMER has stated that the operational year of *Nautile* could be extended to other non-IFREMER scientists if adequate funding is provided, along with an appropriately equipped mother ship staffed by an IFREMER-GENAVIR team. Use of this vehicle may be an alternative when the use of *Alvin* is not possible.

Mir I and *II*

The Russian deep submersibles *Mir I* and *Mir II* have a depth capability of 6,000m. They operate in tandem from the large dedicated research ship *Akademik Mstislav Keldysh*, which employs a side-launch deployment system. The personnel spheres of these vehicles are built from maraged[1]

[1]Maraging is a specific heat treatment process for steel that provides very high strength and durability.

a

b

PLATE 1 Typical human-occupied vehicles: (A) The *Alvin* on a dive. (B) The *Johnson Sea-Link* hovers in the midwater with a pilot and scientific observer onboard.

SOURCE: (A) R. Catanach, used with permission from Woods Hole Oceanographic Institution. (B) Used with permission from Harbor Branch Oceanographic Institution.

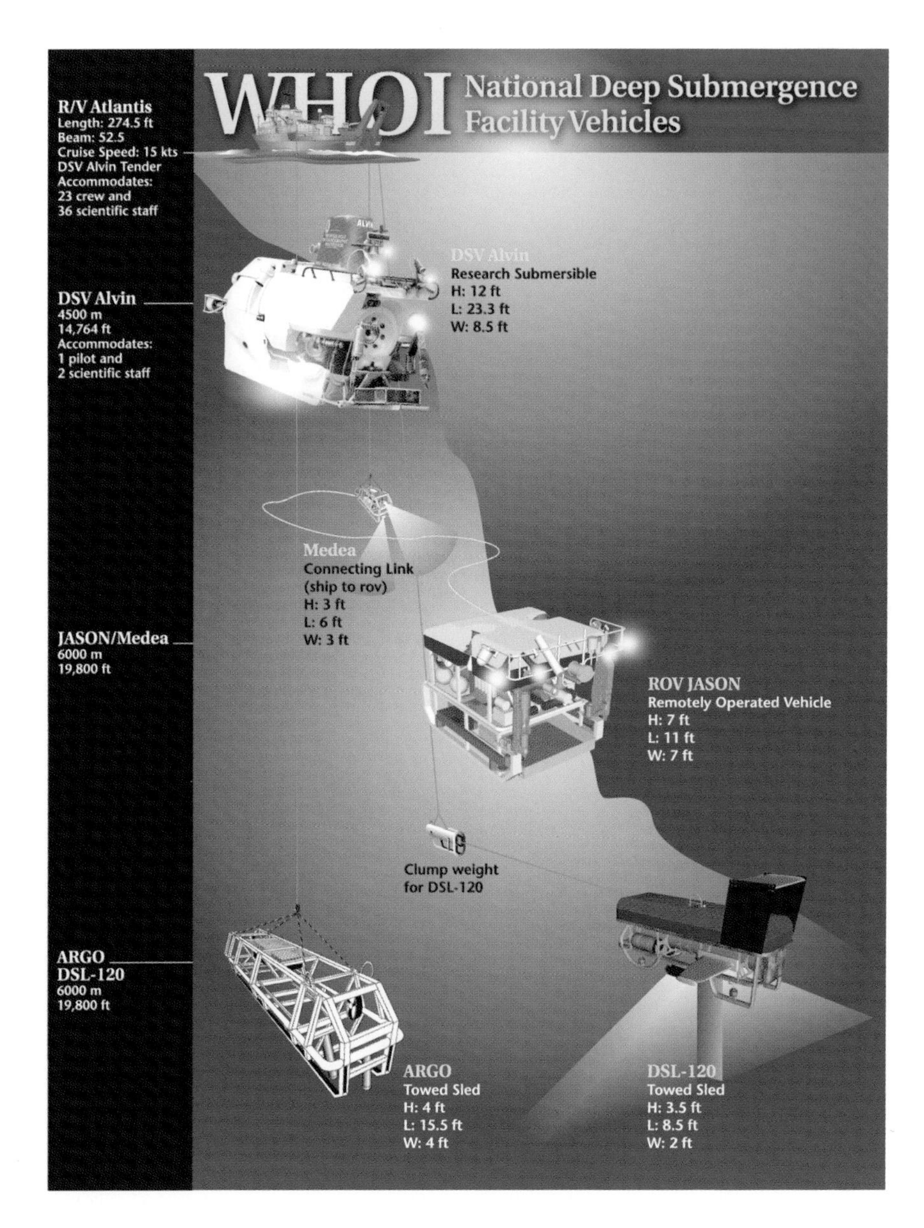

PLATE 2 Vehicles of the National Deep Submergence Facility.
SOURCE: E.P. Oberlander, used with permission from Woods Hole Oceanographic Institution.

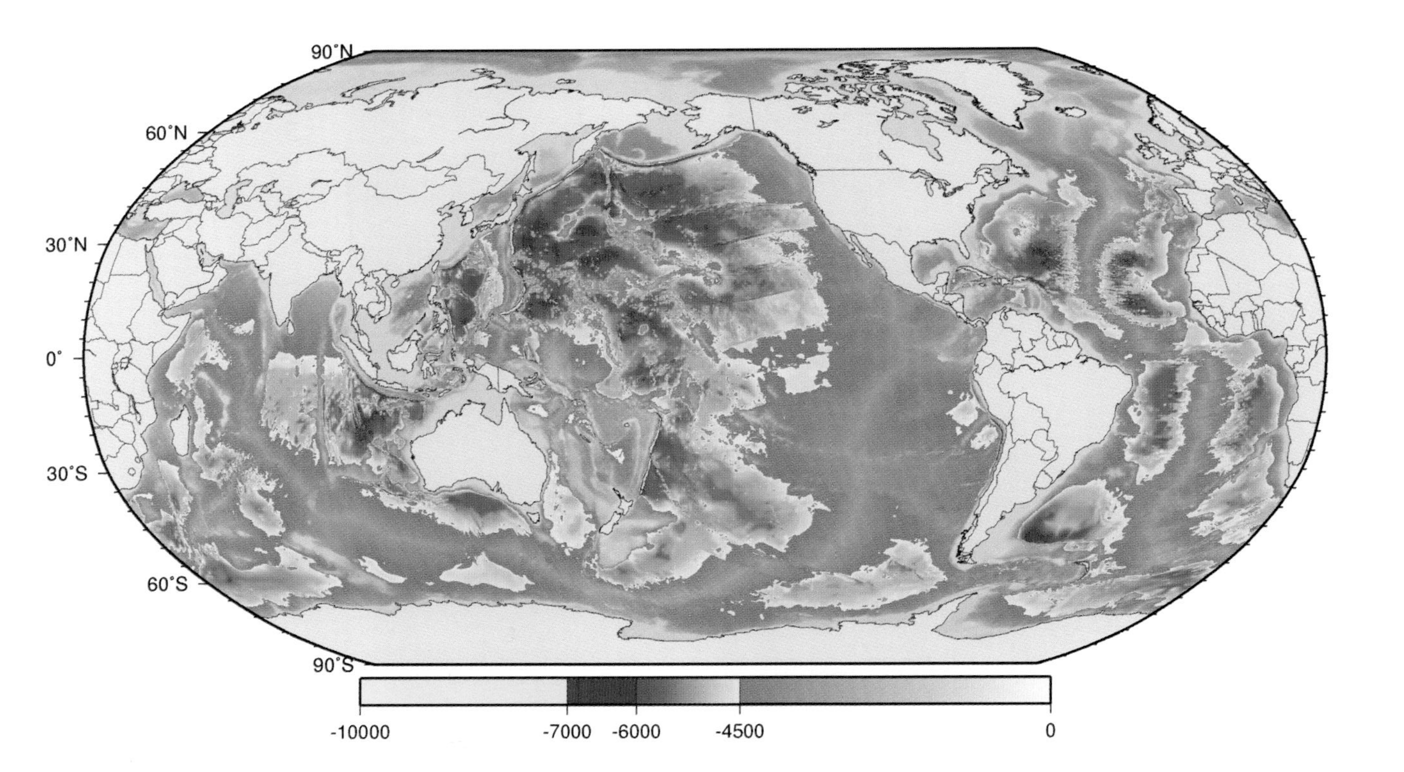

PLATE 3 Graphic representation of the world's ocean basins and their respective depths. *Alvin's* range is shown in gray, with the blue, red, and yellow indicating areas beyond the depth capability of *Alvin*, or deeper than 4,500m.
SOURCE: R. Goldsmith, used with permission from Woods Hole Oceanographic Institution.

a

b

PLATE 4 (A) A black smoker as seen beyond *Alvin's* scientific apparatus for sampling the water chemistry of these chemically rich high-temperature vents. (B) Discovery of hydrothermal vent communities in 1977.
SOURCE: (A) M. Lilley, University of Washington, Seattle; K. Von Damm, University of New Hampshire, Durham, used with permission from Woods Hole Oceanographic Institution. (B) R. Lutz, Rutgers, the State University, Port Norris, N.J.; Stephen Low Productions, Dorval, Canada, used with permission from Woods Hole Oceanographic Institution.

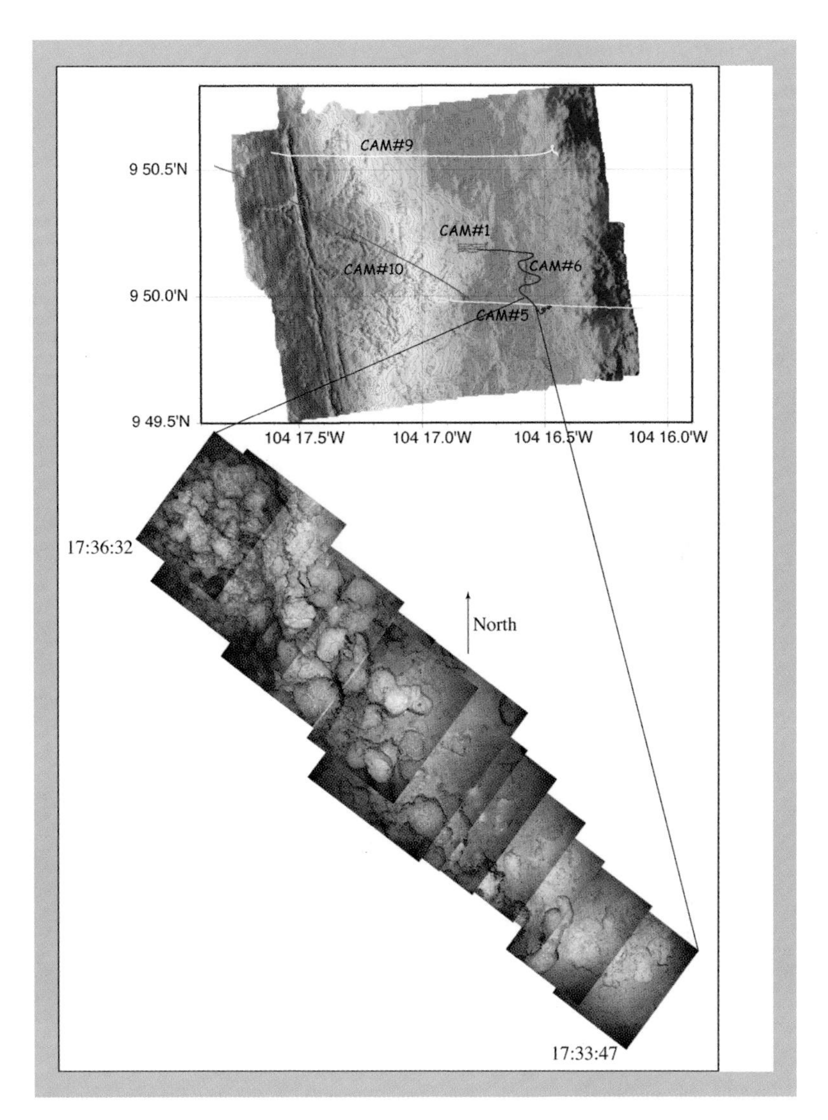

PLATE 5 Example of a contact along one of the scalloped lava flow fronts that are well-imaged in the DSL-120A sidescan data. Mosaic of images is from Camera Tow#6, which traversed the contact several times. Contact is between relatively flat lying sedimented lobate flows (southeast) and a younger lava flow front (northwest) consisting of pillow tubes, bulbous pillows and pillow lavas. The front is 1-2m high, several tens of meters wide and ~100m long. The seafloor behind (west of) the front is relatively flat and consists of relatively unsedimented lobates with collapse pits and an occasional sheet flow channel.
SOURCE: D. Fornari, used with permission from Woods Hole Oceanographic Institution.

a

b

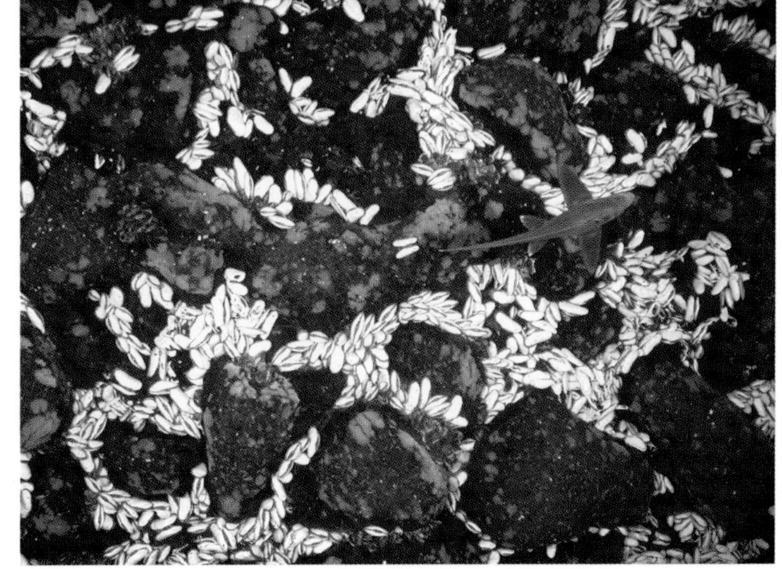

PLATE 6 (A) Discovery of a young vent community during the Galápagos Rift 2002 Expedition, dubbed "Rosebud." (B) Chimera fish swims over numerous large clams and a small clump of mussels that line the cracks between the pillow lavas at the new hydrothermal vent site.
SOURCE: (A) and (B) T. Shank, R. Waller, used with permission from Woods Hole Oceanographic Institution.

a

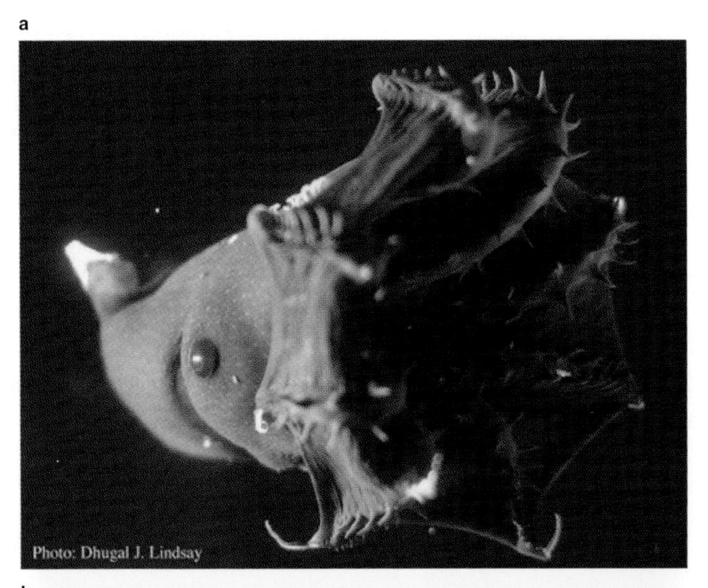

b

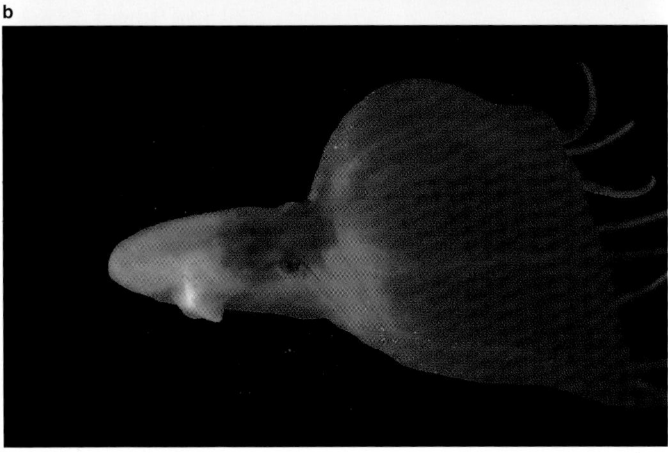

PLATE 7 (A) The cephalopod equivalent to the Coelacanth—*Vampyroteuthis infernalis,* the vampire squid from hell. Captured at 850m depth above Sumisu Seamount, Ogasawara Island Chain, by the Japan Marine Science and Technology Center's ROV *HyperDolphin.* (B) The cirrate octopus *Stauroteuthis syrtensis* is an inhabitant of the narrow benthopelagic transition zone and was known only from very rare net-captured specimens. Recent capture of a live specimen by Harbor Branch Oceanographic Institution's *Johnson Sea-Link* submersible led to the discovery of a remarkable evolutionary transition of suckers into light organs.
SOURCE: (A) Photo taken by D. Lindsay, used with permission from Japan Marine Science and Technology Center. (B) E. Widder, used with permission from Harbor Branch Oceanographic Institution.

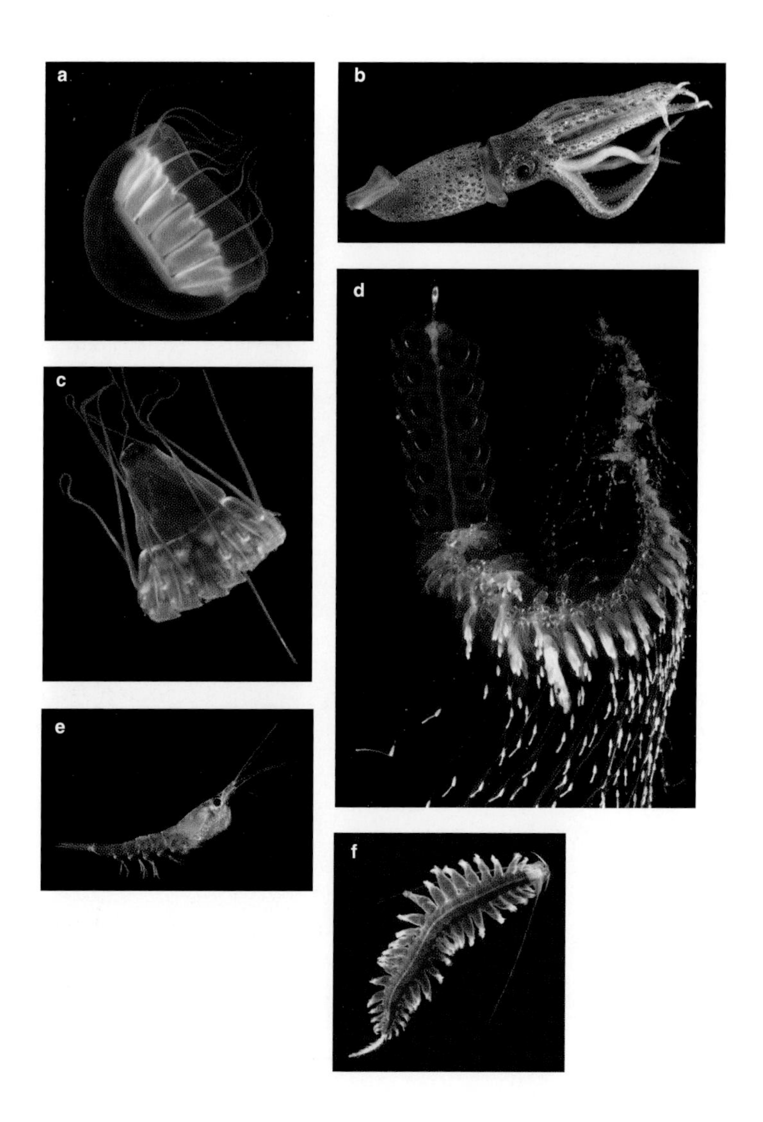

PLATE 8 Typical midwater biota: (A) *Cunina peregrine*, collected at 300m in the Gulf of Maine. (B) *Histioteuthis* sp., called the cock-eyed squid. (C) *Periphylla periphylla*, possibly the most abundant deep-sea scyphomedusa is found in all oceans, generally below 900m. (D) *Nanomia cara* is a common bioluminescent iphonophore in the Gulf of Maine. (E) *Meganyctiphanes norvegica* (Crustacea: Euphausiacea, Northern krill), is critical component of the midwater food web. (F) *Tomopteris* sp., an inhabitant of the midwater, this polychaete worm was collected from 1,200m depth off the Northwest Coast of Africa. The field of view for all figures is no more than 20 centimeters.

SOURCE: E. Widder, used with permission from Harbor Branch Oceanographic Institution.

steel rather than titanium, which allows for greater interior volume at the same outer diameter. The arrangement of viewing ports and the ergonomics inside the sphere of these subs are regarded by users as the best among all research HOVs. Likewise, they have a greater speed and longer underwater duration than most, if not all, comparable systems. They were built in Finland by Rauma-Repola in 1987 and are operated by the P.P. Shirshov Institute of Oceanology of the Russian Academy of Sciences. They have a demonstrated versatility and adaptability and, in addition to their research capabilities, have been utilized for film making, shipwreck survey, and tourist diving. Diving two submersibles simultaneously (they are launched and recovered about 30-40 minutes apart) provides advantages in terms of safety, manipulative operations, scientific observations, and lighting.

The *Mirs* are commercially available to groups other than members of the Russian Academy of Sciences, and requests for their use are made to the Laboratory of Deep Manned Submersibles of the P.P. Shirshov Institute. As with all deep submersible assets from any country, availability is often governed by the availability of the mother ship, transit time or distance, and funding. Because of this, the *Mirs* do not have full schedules and thus are not utilized to their full potential. The *Mirs*, therefore, represent a possible alternative for scientists that cannot make use of NDSF vehicles.

REMOTELY OPERATED VEHICLES

Remotely operated vehicles have improved significantly over the last decade, which has given them wider acceptance in the deep submergence research community. They are capable of diving as deep, and often deeper, than HOVs, and improvements in video and camera technologies have provided a better viewing experience than older models. Commercial enterprises such as the oil and gas industry use them routinely in construction and maintenance of offshore oil platforms and in burying cables. Less expensive to build than HOVs, ROVs are able to perform a variety of scientific tasks that have contributed significantly to the study of oceanography. The following material describes some of the deep-sea ROVs currently used for scientific purposes (also see Table 3-2).

Jason II and *Medea*

The *Jason II-Medea* ROV system is an NDSF asset managed by WHOI. A precision multisensory imaging and sampling platform, *Jason II* is rated to a maximum depth of 6,500m. *Medea* operates with *Jason II*, serving to

TABLE 3-2 Deep-Sea ROVs Used for Scientific Purposes

HOV	Operator	Maximum Operating Depth (m)
Kaiko[a]	JAMSTEC, Japan	11,000
UROV7K	JAMSTEC, Japan	7,000
Jason II-Medea	WHOI, United States	6,500
Victor 6000	IFREMER	7,000
ISIS	Southampton Oceanography Centre, U.K.	6,000
ATV	Scripps Institution of Oceanography	6,000
Ropos	Canadian Scientific Submersible Facility, Sidney, B.C., Canada	5,000
Tiburon	Monterey Bay Aquarium Research Institute (MBARI), United States	4,000
Hyper Dolphin HYSUB 75-3000	JAMSTEC, Japan	3,000
Ventana	MBARI, United States	1,850

[a]The vehicle subunit of the ROV *Kaiko* system was lost on a cruise in May 2003.

provide lighting as well as tether management, which reduces the amount of surface motion influence on the *Jason II's* tether. Both are equipped with high-quality cameras and lighting systems. The system also has several acoustic sensors, a forward drawer, and two swing arms. It has the capacity for high-speed digital link-up and sophisticated sensors to allow a high degree of maneuverability in tight spaces.

Like the *Alvin*, *Jason II-Medea* is an NDSF asset, the use of which must be requested from the UNOLS NDSF. *Jason II* is usually contracted through funding by NSF, NOAA, or the Navy. At present, *Jason II* is scheduled at 100 percent of the available time and, as is discussed below, is projected to be heavily oversubscribed in 2004.

Ventana

The Monterey Bay Aquarium Research Institute (MBARI) in Moss Landing, California, operates the scientific ROV *Ventana*, which is a work-class system built by ISE in Canada and extensively modified for science. It has a depth range of 1,850m and a broad suite of instrumentation, samplers, and tools, including high-definition television (HDTV) and modular tool sleds for benthic and midwater work, as well as for rock drilling, vibra-coring, and cable laying. First deployed in 1988, *Ventana* has made nearly 2,500 working dives and has more than 7,000 hours underwater. It is currently operated chiefly on day cruises out of Moss Landing Harbor. The surface support ship employs a line-of-sight microwave link to shore through which it transmits a live video feed from the ROV to the home laboratory and to the Monterey Bay Aquarium for interactive feedback and outreach.

The *Ventana* has a full schedule each year. It is used principally by researchers at MBARI and its collaborators, which may include co-PIs and scientists from other institutions. MBARI makes 30 days available each year, through the West Coast NOAA Undersea Research Program (NURP) Center at the University of Alaska, for ROV access by non-MBARI scientists.

Tiburon

Tiburon is a second-generation research ROV built at MBARI, with a 4,000-m-depth range. It employs a number of innovative design characteristics that reflect the needs of researchers. These include a variable ballast system that allows the vehicle to be trimmed repeatedly to neutral buoyancy, a camera and light system that slaves cameras and lights to the focal point of a master camera, and a quiet electrical propulsion system. Like *Ventana,* the *Tiburon* has modular tool sleds and can carry a broad range of work packages. *Tiburon* is deployed through a moon pool in the center of a SWATH (small waterplane area twin hull) vessel, which provides a very stable operational platform in advanced sea states. The *Tiburon*'s availability and access are the same as the *Ventana*'s. Like the *Ventana, Tiburon* is used principally by researchers at MBARI and its collaborators, which may include co-PIs and scientists from other institutions. The R/V *Western Flyer* is the support vessel for ROV *Tiburon* and is utilized primarily for extended operations.

Ropos

Ropos is a 5,000-m ROV available on a for-lease basis and operated by the Canadian Scientific Submersible Facility (CSSF), a not-for-profit private sector corporation. This ROV is a "fly-away" system that has been deployed successfully from a number of ships of opportunity. For deep operations, *Ropos* is connected to a cage by a 300-m flying tether. The cage and ROV are launched and recovered as a unit, but at depth the vehicle operates independently within its tether radius. *Ropos* has five-function and seven-function manipulators, two main video cameras, a 36.3-kg payload, and a wide variety of sampling devices and instrumentation. It has more than 3,000 hours in operation compiled during more than 500 dives.

Ropos is an active system that is often booked far in advance. Nevertheless, it is typically idle at various times throughout the year and could be used more often. The CSSF is based at Canada's Institute of Ocean Sciences on Vancouver Island and has working relationships with the University of Washington, NURP, Natural Resources Canada, and several Canadian universities.

HyperDolphin

The JAMSTEC ROV *HyperDolphin* is capable of operations to 3,000m depth. Maximum payload is 100 kg (in air), and it is equipped with two manipulator arms, one with 7 degrees of freedom and the other with 5, and has six hydraulic thrusters. The ROV *HyperDolphin* is equipped with a high-definition camera integrating an ultrasensitive super-High-Gain Avalanche Rushing Photoconductor (HARP) tube and a regular CCD camera on the starboard swinging boom arm. There are five 400-W metal halide lamps, two situated on the port swinging boom arm and one on the starboard swinging boom arm. These arms are usually opened such that the lights optimize the field of view of the high-definition camera, but they can also be moved to optimize lighting during observations of individual organisms in situ. The remaining two lights are forward pointing and are fixed to the frame of the vehicle.

UROV7K

The *UROV7K* is theoretically capable of operating to 7,000-m depth, although it has been tested only to 3,000m. Power is supplied by two lithium ion batteries, and control signals, data streams, and video feeds are communicated via a single microfiber-optic cable of 1-mm diameter. It is currently being modified to operate from the ROV *Kaiko* launcher system through *Kaiko's* secondary tether (allowing power to be supplied from the surface) and twice the number of thrusters.

All deep-sea submersible systems at JAMSTEC (including the *UROV7K* and *Urashima*, discussed below), apart from the recently retired *Shinkai 2000* system, are currently being used to their full potential, and no mechanisms exist to allow the purchase of ship or submersible time. This results from a funding structure whereby JAMSTEC receives payment from the Japanese government for running the facilities, and no provision for external funding exists. Non-Japanese scientists, however, have access to the deep submergence facilities through joint proposals submitted through a primary investigator employed by a Japanese research or educational institution. JAMSTEC has several memoranda of understanding with various international research institutions that also waive onboard costs incurred by scientists at those institutions.

Urashima

The AUV *Urashima*, which can also be operated as an ROV, is capable of operations to 3,500-m depth. Power is supplied by two lithium ion batteries and fuel cells. Control signals, data streams, and video feeds are communicated via a single microfiber-optic cable. Video frame grabs (512 dots, 224 lines) can also be sent acoustically every 8 seconds.

Victor

In 1999 IFREMER launched the *Victor*, which has a depth capability of 6,000m and weighs 4.6 tonnes. It has two manipulators and is presently equipped with a sampling sled designed to work on-site (collect samples and manipulate a number of sensors). There is an ongoing project to build a survey sled equipped with microbathymetry and sonar imagery, wide-angle optical imagery, and possibly other sensors, for high-resolution survey of the seafloor.

Like the *Nautile*, the *Victor* is made available to IFREMER scientists but appears to be underutilized. Although not limited to a six-month operational year, the *Victor* is nonetheless underutilized due to limited funding and the moderate size of the French scientific community. IFREMER has an agreement with a number of countries, including the United States, to exchange ship time and assets. Under this agreement, *Victor* has been successfully used several times on the German *R/V Polar Stern* with the IFREMER-GENAVIR operating team.

ATV

ATV is a large conventional ROV capable of operating at depths approaching 6,000m. *ATV* has hydraulic-powered vertical and horizontal thrusters, two manipulators, lights, video cameras, and the capability to mount additional tools. Built for the U.S. Navy and used in a number of operations including the documentation of the *USS Yorktown* on the floor of the Pacific, the *ATV* was transferred to Scripps and, after about a three-year layup, was operated successfully from *R/V Revelle* off San Diego in May 2003.

AUTONOMOUS UNDERWATER VEHICLES

Autonomous underwater vehicles comprise the third major class of undersea vehicle types. They are, in general, untethered, self-powered, and piloted by a preprogrammed onboard control system. Less mature than HOV and ROV systems, AUV technology is in a phase of rapid growth and expanding diversity, with applications in the military, industry, and research (NRC, 1996; Robison, 2000). Some advantages of AUVs are that they are unencumbered by tethers, allow a surface support vessel to conduct other activities while the AUV is deployed, and can work in regions that are inaccessible or hazardous to HOVs and ROVs (e.g., under ice, high sea states). Disadvantages of AUVs include their limited capability to conduct manipulative tasks or sampling, their inability to respond to unanticipated phenomena or circumstances once launched, and their limited space for a scientific payload. In addition, AUVs are limited by their onboard battery power, which supports relatively short missions. Currently fuel cells and solar cells are in the early stages of development and deployment. Solar cells can potentially recharge batteries when the vehicle surfaces periodically.

A growing number of AUVs are being developed for scientific purposes, many with divergent mission-specific designs and capabilities. Woods Hole Oceanographic Institution's *ABE* (Autonomous Benthic Explorer) (Figure 3-2) is a seafloor survey vehicle with a depth range of 5,000m, docking capability, and control modes for hovering and terrain following (Yoerger et al., 1998). Scripps Institution of Oceanography's *Rover* is a tracked benthic vehicle designed to conduct time-series transects, sampling, imaging, and instrument deployments on the seafloor at depths to 6,000m (Smith et al., 1997). Both *ABE* and *Rover* can be programmed to enter "sleep" modes to conserve power during extended deployments. *Autosub* was developed at the U.K. Southampton Oceanography Centre as a high-endurance, broad-scale survey vehicle with a 1,600-m-depth range and a 100-kg (in water) payload. All three of these vehicles have performed significant and successful scientific missions, including *Autosub*'s research under the Antarctic ice (Brierly et al., 2002).

With increasing levels of sophistication in their instrumentation, distributed network architecture, and programming, AUVs are performing increasingly complex tasks. Modular design for core components and expandable payload bays allow the incorporation of a variety of instrument packages. Improved docking characteristics will enable recharging batteries, adjusting control programs, and downloading data. Ultimately, with improved acoustics and modems it may be possible to put real-time human intervention into the control loop. Appendix C contains a sample listing of currently operable AUVs used by science institutions throughout the world.

FIGURE 3-2 The autonomous underwater vehicle *ABE* being lowered into the sea from aboard a support ship.
SOURCE: D. Fornari, used with permission from Woods Hole Oceanographic Institution.

The scale of investment required for building or purchasing an AUV is much smaller than for ROVs or HOVs, and the operational cost is also much lower. Therefore, AUVs are often owned and operated at the institution level rather than the national level. They are very powerful tools, complementary to HOVs and ROVs, and are often used in association with other assets. Because of their limited payload and power supply, they are generally assigned for specific tasks. Their continuing technological development, however, makes them more and more efficient. Therefore, they undoubtedly should be a component of available assets for deep submergence science.

FIXED OCEAN OBSERVATORIES

Ocean Observatories Initiative

In the last decade, significant elements of the oceanographic research community have embraced a strategy for sustained time-series investigations, primarily to understand temporal variability and causality in Earth and ocean processes. In deep submergence science, this trend is represented by the recent evolution from exploratory deep submergence expeditions toward regular repeat visits to selected sites. In the United States, the overall trend probably will culminate in the very near future in a major investment in long-term seafloor observatory infrastructure by the NSF Ocean Observatories Initiative (OOI). The OOI is the outgrowth of national and international community planning efforts, and various elements of seafloor and ocean observatory science have been addressed in two recent National Research Council reports. The first of these reports, entitled *Illuminating the Hidden Planet: The Future of Seafloor Observatory Science* (NRC, 2000), documented the need for long-term fixed observatory sites in the oceans for conducting basic research to address a broad range of fundamental scientific issues in both ocean and Earth science, and concluded that establishing such observatories is feasible in concept. The second report, entitled *Enabling Ocean Research in the 21st Century: Implementation of a Network of Ocean Observatories* (NRC, 2003a), addressed in more detail the implementation of a seafloor observatory network for multidisciplinary ocean research in the context of the pending NSF OOI and specifically examined the impact on both the UNOLS fleet and the pool of deep submergence assets in the research community. These observatories, if constructed, will have a significant impact on deep ocean research, especially for time-series studies. The deep submergence vehicles (DSVs) needed to support OOI are separate from, and would be in addition to, those discussed in this report.

THE NATIONAL DEEP SUBMERGENCE FACILITY

As discussed in Chapter 1, the NDSF was created in 1974 by the National Oceanic and Atmospheric Administration, the Office of Naval Research, and the National Science Foundation to provide the nation with a core operational deep submergence team. The first 21 years of NDSF operations included only HOV work with *Alvin*. During this period, various towed geophysical packages and ROVs were developed at a number of U.S. oceanographic institutions, including WHOI. In 1995, the NDSF collection of submergence assets was expanded to include some of the tethered vehicles that had been developed at WHOI, including the ROV system *Jason II-Medea* and the *Argo* towed camera system. Assets currently provided by the NDSF include the HOV *Alvin*, the ROV system *Jason II-Medea*, the *Argo-II* towed camera system (Figure 3-3 and 3-4), and the DSL-120A side-scan sonar system. Use statistics for NDSF vehicles are summarized in Figure 3-1.

It is generally agreed that the stable, ongoing support for NDSF has been beneficial in providing this core submergence capability to the U.S. research community. NDSF platforms support a variety of science mis-

FIGURE 3-3 *Jason II* is lowered by crane into the sea.
SOURCE: D. Fornari, used with permission from Woods Hole Oceanographic Institution.

sions, as discussed in Chapter 2. The increasing diversity of tasks associated with an ever-expanding suite of science missions will serve only to increase pressure on the NDSF to expand its assets. For a variety of reasons, including vehicle support and the nature of current funding schemes for DSV use, NDSF vehicles are the most often used assets for deep submergence research in the United States. The accessibility of these vehicles can therefore be a limiting factor in the growth of deep ocean research.

PATTERNS OF USE

At the end of 2002, *Alvin* had accomplished a total of 3,859 dives (Table 3-3). It is one of the most utilized deep-sea vehicles in the world and has a remarkable success record. *Jason* was built in the late 1980s, and now, with the more sophisticated *Jason II* (available since July 2002), the demand has started to grow. For the next two years, *Jason II* will be used more than *Alvin*.

Since the launch of *Alvin*, *Jason*, and *Jason II* the community of users has evolved to include a broad array of scientific disciplines including geology and geophysics, chemistry, physical oceanography, biology,

FIGURE 3-4 The National Deep Submergence Facility towed camera system *Argo II* preparing for deployment behind its mother ship.
SOURCE: H. Sulanowska, used with permission from Woods Hole Oceanographic Institution.

and engineering. Although the science conducted on any of the *Alvin* and *Jason II* dives can be multidisciplinary, Figures 3-5 and 3-6 break down the operational days for *Alvin* and *Jason II*, respectively, into the predominant science done for each dive based on individual dive logs. A few main points should be noted based on these data. First, *Alvin* and *Jason II* are used predominantly for marine geology and geophysics

TABLE 3-3 A*lvin* Statistics, 1964-2002

Total dives	3,859
Average depth per dive	2,303 meters
Total time submerged	26,503 hours
Average time submerged per dive	6.87 hours
Average bottom time, 1997-2002	5 hours
Total persons carried	11,570
Total of individuals carried	7,287

SOURCE: Data from R. Pittenger, Woods Hole Oceanographic Institution, written communication, 2003.

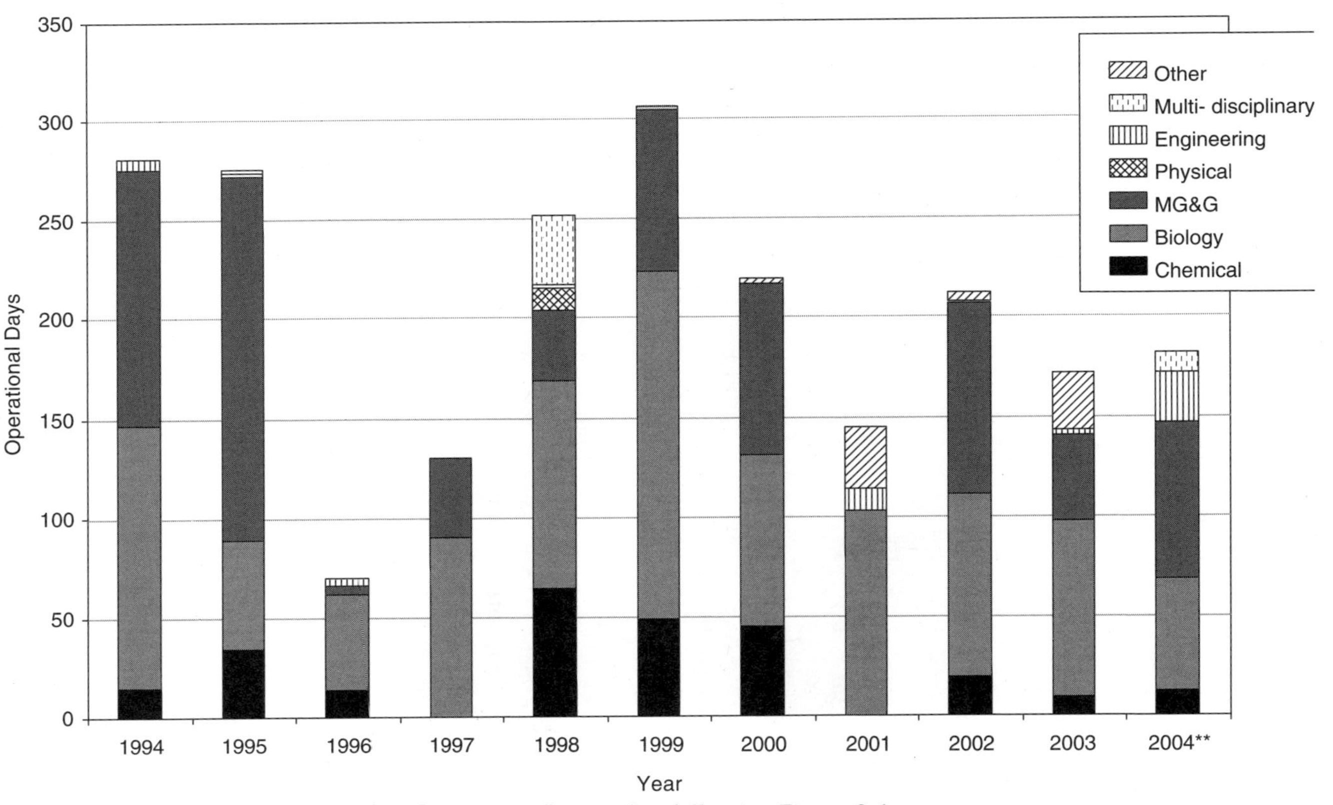

FIGURE 3-5 Total *Alvin* operating days by science. See caption following Figure 3-6.
SOURCE: Data from A. DeSilva, University-National Oceanographic Laboratory System.

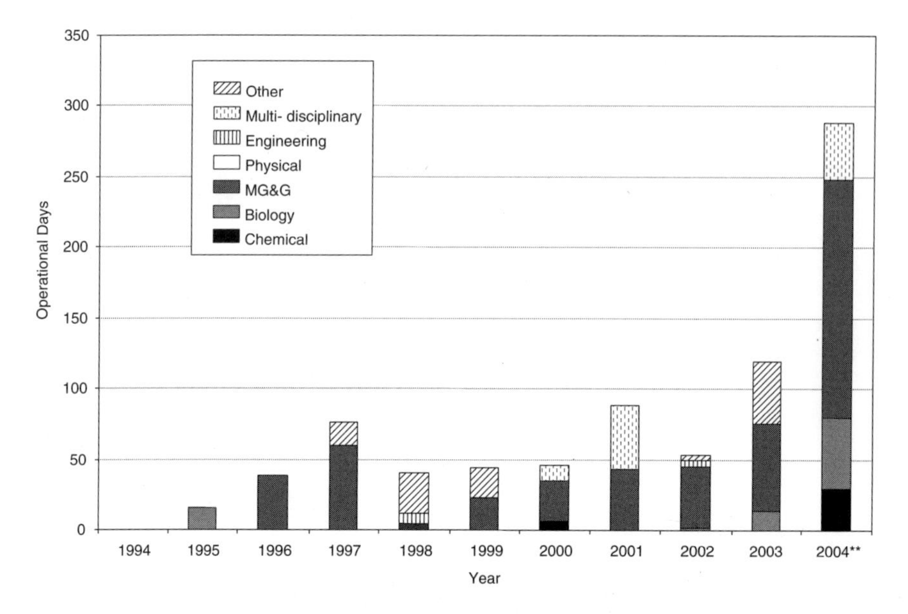

FIGURE 3-6 Total *Jason* operating days by science. The statistics for Figures 3-5 and 3-6 are based on ship utilization data forms and *Alvin* dive logs. Many cruises are multidisciplinary in nature. In order to break down a cruise by discipline it was often necessary to refer to the individual dive logs. An operating day represents all chargeable days away from home port. If a cruise included use of multiple vehicles (*Alvin, Jason, Argo II*, DSL) the cruise operating days were divided proportionally. The statistics for 2004 are based on ship time requests and ship scheduling information. They represent total operating days corresponding to each funded request. Currently there is more funded work than can be scheduled in 2004 on UNOLS facilities. Some programs will be deferred until 2005.
SOURCE: Data from A. DeSilva, University-National Oceanographic Laboratory System.

(MG&G), and biological research, with *Jason II* use moving more toward the MG&G disciplines and very little biology. Second, *Alvin*, which was traditionally used mostly for MG&G, has experienced a general increase in biology users with a corresponding drop in use by the MG&G community. Finally, with the introduction of *Jason II* in the summer of 2002, there has been a dramatic increase in the number of planned dives (based on ship time requests and ship scheduling information) for 2004. It is unclear whether the recent association of *Jason II* with MG&G cruises is a product of scheduling practicalities or a reflection of changing preferences. These data may also illustrate an increase in biological efforts as a consequence of the serendipitous discovery of deep ocean communities by geology, geochemistry, and geophysics researchers using HOVs. Information on their environmental aspects, hydrothermal circulation, and structure is needed by biologists in a setting that is enhanced by other research efforts. The MG&G community, which in effect does reconnaissance work for the biology community, may be using ROVs more often than HOVs simply because they are more available.

Geographic

Alvin is supported by the surface ship, R/V *Atlantis*, the only surface vessel equipped with an adapted A-frame capable of recovering *Alvin*. Over the past two decades the ridge-crest time-series studies at particular locations have dominated the use of *Alvin* as well as various NDSF and non-NDSF tethered vehicles. The prime sites for these studies are at the Juan de Fuca Ridge, the East Pacific Rise north of the equator, and the northern Mid-Atlantic Ridge. The strong pressure to revisit these sites on an annual or biannual basis has been the major factor limiting access of *Alvin* to other geographic locations in the oceans. This has been termed the "yo-yo" effect as *Atlantis* and *Alvin* have been pulled back and forth through the Panama Canal to and from these sites at the appropriate weather windows. While the science generated at these sites is acknowledged to be very strong and the overall pattern is well justified by strong proposal pressure, the net result has been a geographic limitation on HOV use elsewhere in the oceans.

ROVs are more easily transported and therefore can be used on various ships provided that they have sufficient deck space, a suitable location for launch and recovery, and dynamic positioning (DP) capability. The recent availability of the ROV *Jason* and then *Jason II* has demonstrably altered the patterns and nature of deep submergence research sites by expanding the number of support ships available, thus eliminating the need for a single ship dedicated to a specific submersible. Increasing requests for *Jason II* demonstrated the demand for access to submersibles in other geographic areas of great interest to scientists such as the western Pacific, where the deepest trenches are located, or the Indian Ocean, which is crucial to understanding the distribution of biogeographical provinces at mid-ocean ridges. Originally envisioned as a complement to *Alvin*, *Jason* has demonstrated the unique value of ROV platforms for deep ocean research.

The mother ship of both the *Nautile* and the *Victor* is presently the *R/V Atalante*, and a new submersible carrier the *Pourquoi pas?*, is under construction and should be finished by 2005. However, it will be shared with the French Navy, which will bring some logistical constraints. The *Victor* can also be used on the *Thalassa*, which is tied to the North Atlantic. Under the frame of bilateral agreements, it has also been deployed from the *R/V Polar Stern*. The call for proposals each year generally includes the planned location of the ships, which implies where the submersibles can be used.

In Japan the submersibles have to go back to Japan every year. The Japanese submersible fleet operates mainly in waters around Japan and the western Pacific, but surveys have also been conducted, sometimes routinely, in other areas such as waters off Hawaii, Papua New Guinea, Java, Micronesia, Fiji, the eastern Pacific, the Indian Ocean, and one cruise even to the Atlantic Ocean. Consequently, they essentially work around Japan and in the western Pacific (with some cruises in waters off Hawaii, Papua New Guinea, and Guam), although in 1998, the *Shinkai 6500* had an around-the-world trip diving in the Atlantic and Indian Oceans. The Russian *Mirs* are leased most of the time and have no geographical constraints. They are operated from a dedicated mother ship, the *R/V Akademik Keldysh*. The *Mirs* have conducted operations in all of the major oceans except the Southern Ocean. The *Keldysh* is one of the largest oceanographic research ships in the world and is capable of very long cruises.

Water Depth

The depth range of *Alvin* is presently limited to 4,500m. Opportunities to reach this depth however, have not often been exploited (Table 3-4). This may be because most scientific cruises focus on ridges, which are located at a 1,500 to 3,500m depth range. Moreover, logistically the shallow portions of trench systems do not fit in the yo-yo pattern in which the *Alvin* is caught. The *Nautile* also rarely reaches its 6,000m depth range and, in fact, did not dive deeper than 5,000m in 2000 and 2002. Even so, the scientific return on these dives has been quite exciting with the exploration of key areas such as Hess Deep or Pito Deep in the Pacific and deep transform faults such as Vema, Kane, and Saint Paul in the Atlantic. The *Shinkai 6500* goes deeper than 5,000m on more of its dives, possibly due to the proximity of deep trenches and the focus of the Japanese scientific community.

Time Series

Over the last decade, time-series studies of key sites of interest have played an increasing role in ocean science. Much of the use of deep submergence assets has been devoted to this important time-series work, which may be addressed in the future by fixed ocean observatories. As discussed earlier, this study draws on two recent NRC reports *Enabling Ocean Research in the 21st Century: Implementation of a Network of Ocean Observatories* (2003a) and *Exploration of the Seas: Voyage into the Unknown* (2003b). *Enabling Ocean Research in the 21st Century* and *Exploration of the Sea* each propose a suite of vehicles to support those endeavors. The recommendations

TABLE 3-4 Comparison of Maximum Depths Reached by the *Alvin*, *Nautile*, and *Shinkai 6500*

Submersible		1990	1991	1992	1993	1994	1995	1996	1997	1998	1999	2000	2001	2002
Alvin														
	Dives >4,000m					27	13		2		8		4	1
	Total dives	154	152	76	136	181	172	49	97	134	188	138	91	115
	>4,000m (%)	0	0	0	0	15	8	0	2	0	4	0	4	1
Nautile														
	Dives >5,000m	3	12	6	5		7	2	1	2	12			
	Total dives	66	128	114	106	84	86	114	57	113	104	40	0	35
	>5,000m (%)	5	9	5	5	0	8	2	2	2	12	0	0	0
Shinkai 6500														
	Dives >5,000m	4	6	24	14	6	14	14	6	9	12	13	8	4
	Total dives	36	64	61	35	64	48	51	46	63	61	66	48	85
	>5,000m (%)	11	9	39	40	9	29	27	13	14	20	20	17	5

SOURCE: Data from R. Pittenger, Woods Hole Oceanographic Institution, Woods Hole, MA, written communication, 2003.

in this report are above and beyond any capabilities called for in those two reports. This report does not revisit the scientific justification in detail here; the committee concurs with recommendations put forward in both reports. Specifically, the NRC (2003a) report concluded that ROVs would be the primary deep submergence vehicles needed by the OOI and that about two ship-years of ROV time would be required annually by the OOI. It recommended that when the OOI is implemented, NSF should provide submergence support for it by adding a second deepwater ROV to the NDSF pool and provide academic access to a third ROV that is not necessarily within the NDSF. This study assumes that these recommendations will be followed when the investment is made in the OOI. In other words, recommendations made in this current study for the mix of deep submergence assets required for future scientific needs, exclusive of the OOI, are independent of, and in addition to, those made in the NRC (2003a) study on implementing an ocean observatory network.

Pertinent to the NRC (2003a) study, key findings and recommendations: "ROVs are anticipated to be the work-horses of deep-ocean observatories. ROV resources will be needed for installation of seafloor observatories, connecting moorings to seafloor junction boxes, installing experiments, and servicing or repairing instruments and network equipment on the seafloor. Their durability on the bottom, heavy-lift capability, and high available power make them indispensable assets for observatory operations. . . . Human-occupied vehicles (HOVs) are not likely to play a major role in routine observatory installation and servicing due to their lack of power, short dive duration, lack of heavy-lift capability, avoidance of suspended cables, and limited communication capabilities to the surface. However, because HOVs are untethered to the surface and thus highly maneuverable, they may be useful in some instances for conducting scientific investigations around observatory sites, and for initially establishing experiments and locating sensors in areas of complex topography such as a hydrothermal vent field."

The improved *Alvin* replacement recommended in this study can probably accommodate the modest observatory needs for HOVs. It must be emphasized, however, that the additional deepwater ROV recommended in this study is in addition to the ROVs needed for virtual full-time support of the OOI.

DEMAND

For the next two years (2004 and 2005), funding and pending cruises in the United States demonstrate the new interest of the scientific community in using *Jason II* for different reasons. First, the community is convinced of the efficiency of this vehicle to conduct the scientific programs.

Second, it increases the depth range and therefore opens new scientific targets. Finally, location maps clearly show that it breaks the yo-yo pattern that the *Alvin* still experiences. For the first time, the demand on *Jason II* exceeds the demand on *Alvin*.

CAPABILITIES NEEDED TO REACH SCIENCE GOALS

The use of deep-sea HOVs and ROVs allows real-time direct observation (either in person or remotely), mapping, in situ sampling, and experimentation, along with monitoring of the seafloor and water column, and their associated biospheres, as well as their physical and chemical attributes. It is an understatement to say that a number of exciting discoveries would never have occurred if these facilities had not been available. The 40 years of deep submergence science clearly show a trend in initial exploration to repeat measurements and experimentation. A variety of tools have been developed around the vehicles: sampling devices, and physical, chemical, and biological sensors. The image recording capacities have improved dramatically with new camera systems and lights. Currently, it is obvious that the scientific community needs to have access to a full range of assets that can be used in various combinations. HOVs, ROVs, and AUVs each have their own advantages depending on the scientific problem that is to be addressed and the appropriate research protocol. While ROVs and AUVs will undoubtedly become more sophisticated, which may supplant the need for human scientists to carry out deep ocean research directly in many instances, the added value of human perspectives will remain significant. As technology improves to link human eye movements (rather than the hands) directly to automated camera control, it will ensure that HOV and HOV-ROV hybrid systems are superior to ROV systems for a wide range of applications well into the future.

Viewing and Documenting

The primary goal that justifies the use of deep-sea submersible for geologists is to be able to actually view the study site. It is only the combination of mapping, viewing the terrains, and sampling that allows the construction of geological maps that portray the character and extent of lithologic units, as well as their spatial relationships. Viewing also allows the description and measurement of tectonic features, such as orientation of fissures and fault planes, the nature of tectonized surfaces, and easier location of the distribution of hot vents and cold seeps with respect to tectonic activity. This is the normal, necessary approach on land, and only the use of submersibles makes it possible at the seafloor. For biologists, viewing gives them the ability to identify animals through visual cues, observe their behavior, quantify them, relate their distributions to environmental parameters, and sample an undamaged state.

With HOVs and ROVs the ability to view the environment outside the vehicle is essential to conduct a dive; real-time decisions are made on the basis of what is seen. Viewing is achieved directly with the human eye and video cameras in HOVs and through a combination of video cameras with ROVs. In both cases, light is required to illuminate the field of view or water mass, which is completely dark at depths greater than a few hundred meters. Red light illumination has proved invaluable for observing animal behavior unmodified by the effects of light, since most deep-sea animals are sensitive only to light in the blue-green end of the spectrum. With an HOV the human eye has several advantages over cameras. This is particularly true when the lights are extinguished to observe bioluminescence. Although intensified cameras are much improved over the low-resolution, high-noise, long-lag-time system in use a few years ago, there is nothing better than the dark-adapted human eye for observing in low-light underwater conditions. The human eye also has many advantages with the lights turned on. For example, even if the portholes are small because of the pressure, the effective field of view using the human eye is much wider because it can easily follow and focus on objects outside the HOV. This is especially true in the midwater zone where animals move in three-dimensional space, often along different vectors to the movement of the submersible. In order to correctly identify, quantify, and record the behavior of midwater animals, an imaging system must be able to focus rapidly on multiple targets in succession in a three-dimensional space while keeping peripheral vision open to identify alternate targets for subsequent or simultaneous visualization. The major drawback to the human eye is that it does not leave a permanent record for later referral other than that which can be written down or recorded as an audio signal. It also does not allow sharing of visual information with other specialists or the public directly as a video stream would, and numerical data cannot be embedded and synchronized with events efficiently. External cameras on pan-tilt mechanisms can follow moving organisms, but the movement is not instantaneous and focusing is also not as rapid as with the human eye. Several experts commented at open sessions of the Committee on Future Needs in Deep Submergence Science that some of the discoveries made while diving might never have happened using an ROV. The lack of three-dimensional vision on video screens presents problems with the interpretation of the images collected by an ROV. This will certainly be improved in the future with stereo viewing or three-dimensional visualization. The configuration of portholes in *Alvin*, however, should be improved, if at all possible. In the present setup there is minimal overlap between the fields of view of the pilot and the two observers. This makes interaction difficult particularly during manipulations.

At this point it seems that the interpretation of video images on monitors is more straightforward for scientists with HOV dive experience. On the other hand, the monitors in a control room can be watched in real time by a group of scientists who can share their experience. With satellite communications advice can even be requested in real time from specialists in their office on land. JAMSTEC uses a single fiber-optic cable communication system in the battery-powered deep-diving submersible platforms *UROV7K* and AUV *Urashima*. Such a system could also be incorporated into any HOV platform, expanding the number of scientists viewing the deep ocean when desired.

Both HOVs and ROVs record images on video and still cameras. The quantity of data collected tends to increase drastically and consequently raises the question of storage, archiving, and accessibility. Database systems for the storage of nonnumerical data have been developed at institutes such as the Monterey Bay Aquarium Research Institute.

Because of the lack of real-time communication, AUVs are suited only for preprogrammed systematic surveys. The optical images collected allow the construction of image mosaics, which are of great value in documenting an area at a local scale.

Sampling

Sampling in situ, in a well-characterized environment is a major advantage of submersibles. For geologists, the value of collecting a rock from a previously defined and characterized unit is enormous, compared to dredging loose rocks over a distance of several tens to hundreds of meters. The collection of hydrothermal fluids from vents requires placing a syringe or a sampling tube into a hydrothermal conduit that may not exceed a few centimeters in diameter. Delicate biological samples can be collected only with manipulators and must be stored in dedicated boxes immediately. Sampling of some of the fragile gelatinous animals from the midwater requires special tools, and because a failed sampling attempt can destroy the organism that is being sampled, depth perception is extremely important. A trained ROV pilot can use multiple cameras set at different angles to visualize a three-dimensional space, but because this system is not intuitive and sampling efficiency is usually less than with HOVs, the precision of the sampling drastically influences the quality of the sample. The sampling of biological specimens also benefits from a characterization of the environment, that is, the local temperature and fluid chemistry.

Both HOVs and ROVs are equipped with precision manipulators that are operated by the pilot either on the HOV or from the control room on the ship. They allow the pilot to collect rocks from the seafloor (if they are reasonably loose) and also to operate a variety of sampling tools attached to the vehicles, such as water bottles, pumps, slurp guns, sediment corers, rocks drills, and so on. The samples collected are stored directly in a basket or in specially designed containers. This is crucial for an efficient sampling program, and both types of vehicles have been used successfuly during a large number of cruises. In practice, however, manipulations seem to be less efficient through remote control with an ROV because of the lack of three-dimensional vision on video monitors. This is compensated for by the absence of a limitation on the time spent over the seafloor with an ROV. The depth perception provided by direct observation also gives the scientist a better understanding of the context in which samples or data were collected.

Another difference between HOVs and ROVs comes from the combined payload and duration of the dive. Systematic sampling programs can offer unique challenges to deep-water operations. These sampling programs can rapidly meet the weight limit as the collection basket and containers are filled. Moreover, biological samples, displaced from their environment, have to be recovered as soon as possible to avoid deterioration. With HOVs the total duration of a dive rarely exceeds 8 to 10 hours, and the time actually spent over the seafloor is typically 6 to 8 hours. The accumulation of samples collected during this time can normally be accommodated. The samples are recovered every day with the vehicle and are readily accessible for possible experiments onboard. With ROVs the duration of the dive may reach several days. In that case the upper weight limit is reached rapidly and continued sampling requires the use of an elevator. This is essentially a big container with drawers, which is launched directly from the ship and lands on the seafloor. The drawers are filled with samples using the manipulators, and then a release system sends the elevator back to the surface where it is recovered by the ship. Although it is an efficient system, the use of an elevator is time-consuming because the ship and vehicle are committed during the launch and recovery. In acceptable sea states however, the elevators can be recovered by a zodiac or similar small boat. If necessary, an elevator can also be used with an HOV, although no such system currently exists for recovery of samples from within the water column.

Because of the lack of real-time communication and limited payload and availability of energy, AUVs are not appropriate for sampling with manipulators. Simple sampling procedures can be programmed, such as the filling of water bottles at regular time intervals, but AUVs are not likely to become good sampling vehicles in the near future.

Platform Stability

Stability of submersible platforms relative to position above the bottom or within a given water mass is best attained using a variable ballast system. Such systems can be installed on any platform, including HOVs, ROVs and AUVs. They ensure greater visibility and more efficient use of dive time by making the submersible neutrally buoyant at depth without the use of thrusters. This minimizes sediment or flow field disturbance; enables fine manipulation of fragile equipment, samples, or experiments even in regions of rough topography; and allows quantification of objects on the bottom through the ability to keep a stable altitude and in the water column through being able to adjust the speed of forward movement relevant to suspended particles. Near-perfect trim to neutral buoyancy, especially with HOVs and AUVs that are completely decoupled from the surface, also permits implementation of sensitive optical experiments such as low-light-level measurements, which otherwise would be compromised by unwanted stimulation of bioluminescence, and holographic particle velocimetry, which would be compromised by unwanted platform movement and disturbance of natural flow fields. With the notable exception of the *Tiburon*, the current suite of scientific ROVs lacks a sufficiently robust variable ballast system to allow the ROV to become truly neutrally buoyant.

Mapping and Surveying

The resolution of bathymetric and geophysical surveys depends on the distance to the object. Closer distance clearly allows better resolution. For instance, faults with a trough of a few meters are not detected from surface mapping although they play a crucial role in the location of fluid vents and therefore the distribution of associated biology. Magnetization lows associated with hydrothermal circulation were detected only by near-bottom surveying; they now provide a new tool to search for discharge zones on the seafloor. Ideally, the characterization of an area requires a combination of scales from the general picture acquired from surface mapping to detailed mapping using submersibles in selected, particularly interesting, detailed areas. For example, the detailed map of a hydrothermal field, with the location of all the vents with respect to local tectonic features, has to be placed in the general context of the ridge segment.

Although HOVs have occasionally been used for specific surveys such as magnetics, in situ gravity measurements on the seafloor, or routine recording of temperature and some chemical parameters, ROVs and AUVs are clearly better suited for that type of work. Mapping and surveying typically include bathymetry, sonar imaging, wide-angle optical imaging, magnetics, temperature, and salinity. More sophisticated sensors can also be attached, such as nephelometers to record the turbidity of water or manganese sniffers to detect chemical anomalies in the water column. Methane sensors, still under development, will be useful tools to locate regions of active serpentinization or cold seeps in margin environments. Surveys can be conducted at various heights above the seafloor depending on the scientific objectives. Typically, a bathymetric map will have a horizontal resolution of a few meters if collected at 20 to 30 m, and <1 m if collected 8 to 10 m above the seafloor.

ROVs have the advantage over AUV systems of returning data to the ship in real time, allowing the operators to make decisions on the vehicle track. This capability is essential to follow structures such as active faults or plumes in the water column. The ROV is linked to the ship however, and therefore the ship time is completely devoted to the survey. AUVs can be programmed to conduct a systematic survey and have the advantage of being completely autonomous. Once launched, they conduct their own program independently while the ship can be used for other purposes.

Experiments and Time-Series Investigations

Studies of processes and the sampling that occurs at dynamic interfaces, such as hydrothermal plumes (including their propagation and mixing along isopycnals), cold seeps, benthic boundary layers, or seafloor gas hydrates, are best done by submersibles. The intricacies involved in measuring the physical, chemical, and biological properties, and especially collecting samples, from these few-centimeter to millimeter-thick boundaries require specially designed tools that must be handled with sophisticated submersible manipulators. In addition, conducting challenging in situ interactive perturbation experiments and observing the responses to them, including how fast they return to their "steady state," would enhance our understanding of how these systems respond to natural environmental perturbations.

The development and testing of new and novel tools for key in situ measurements, such as instruments for measuring pH and dissolved sulfide content of pristine hydrothermal fluids, or using nuclear magnetic resonance (NMR) to document the growth habit of methane hydrate in sediments, its relationship to hydraulic permeability, and the subsequent impact on slope stability will require frequent access to various submersible vehicles.

On the other hand, the progression from spatial exploration in ocean sciences toward temporal understanding of active processes has played a significant role in the history of the utilization of deep submergence assets, both international and national. This has been manifested in a trend toward repeat visits at selected sites for time-series sampling of active processes, with irregular sampling intervals owing to the complexities of the scientific proposal and to ship and submersible scheduling procedures. This trend has contributed to the development of the OOI and may result in the establishment of long-term observatories at a few of these sites. Prime examples are the time-series investigations of hydrothermal, magmatic, tectonic processes at representative ridge-crest sites, some of which are now designated as RIDGE 2000 "Intensive Studies" sites. A similar theoretical and practical approach is emerging in deep submergence investigations of active processes in other benthic environments, for example, subduction settings and sites of significant gas hydrate accumulations.

The diverse scientific goals and the large depth range in margins from slopes to trenches, and in pelagic environments from ridges to basins, require an appropriate array of ROVs, HOVs, and AUVs, from shallow depths to ~7,000m, to reach most trenches. In convergent margins with deep trenches (water depths >4,000m), ROVs that are coordinated with AUVs currently seem preferable because of the limited bottom time provided by HOVs at such depth. Some areas, however, have sites and seeps with communities of organisms that would benefit from direct observation and delicate sampling, which may necessitate HOV use. Future opportunities and benefits to interface with moored instrumented observatories and with instrumented boreholes, for basic scientific objectives and novel experiments, will have to be considered when new vehicles are designed. To date, most of the data downloading from monitoring instruments in boreholes, as well as some new instrument deployments, has been carried out with submersibles.

The ability to respond rapidly to significant transient events such as large earthquakes, landslides, tsunamis, volcanic eruptions, phytoplankton blooms, eddy formations, and transient upwelling events, and to observe, sample, and interact at the sites of these events, is essential. These events are an integral part of the Earth energy cycle, which involves production, transport, recycling, the aging and subduction of the oceanic crust, and the origin of continents.

Outreach

Deep-sea exploration and discovery are inherently exciting and in many respects they are ideal topics to share with the public. The means for sharing images and data and for real-time participation in deep-sea research activities are readily available. A microwave link from surface vessels to the shore allows visitors at the Monterey Bay Aquarium each day to interact with research scientists via live video broadcast from deep-working ROVs.

Likewise, the Jason Project, the BBC (British Broadcasting Corporation), and JAMSTEC have used satellite links to carry ROV-generated video live to distant viewers. With ROVs the transmission of information and images is a fundamental aspect of their operation, connecting not only to pilots and scientists topside but also to students and the public.

Until recently, live transmission from HOVs at depth was limited chiefly to through-water acoustics with a bandwidth too narrow for anything but voice and position data, and low-resolution frame grab images. The principal problem with acoustic transmission to the surface is stratification of the water column. One solution is to transmit to an acoustic receiver (modem) at depth that is connected by wire or optical fiber to the surface. Even more promising is the use of an optical fiber that links the HOV to the surface. This fiber would provide for a direct, high-bandwidth link to the surface without the problems of acoustic transmission, and with reduced drag and handling problems compared to thicker ROV tethers.

High-bandwidth communications from HOVs will improve their scientific productivity by involving many more researchers in the real-time experience, just as ROVs do. Similarly, AUVs will benefit from real-time human presence in the control loop. With both kinds of vehicles the potential for outreach to the public is also greatly enhanced. These advances, coupled with rapidly evolving knowledge base systems, comprise an expanding wave of outreach applications involving all three classes of deep-diving vehicles.

CONCLUSION

Each type of deep-sea vehicle definitely has its own advantages and disadvantages. The best approach to deep submergence science involves a nested survey strategy that utilizes a combination of tools in sequence (possibly over multiple expeditions) for investigations at increasingly finer scales. For geological applications, initial reconnaissance surveys are best achieved using ship-based swath-mapping systems and tethered survey vehicles (including ROVs and AUVs). After the local context is established by such surveys, detailed descriptions of specific sites or work in the water column can benefit from the direct experience afforded by the use of HOVs. Experiments and observatory work that require longer time on the seafloor and/or heavy lift capabilities on already-characterized sites are best conducted with ROVs.

At present, the DSVs of NDSF, specifically *Alvin* and *Jason*, are the most widely used DSVs available to deep ocean scientists in the United States. While this demand is undoubtedly a testament to the scientific support provided by NDSF and the reliability of these platforms, it also represents the growing interest in conducting research into unique and interesting processes that can best be studied through the use of deep submergence assets. As discussed earlier, recent trends in total demand for access to *Alvin* have stabilized to the point where *Alvin* is essentially fully subscribed each year (Figure 3-5). In addition, improvements in the design of *Jason II* have greatly enhanced the usefulness of this ROV for deep submergence work, which has resulted in an increased demand for access to this platform. As a consequence, *Jason II* is now significantly oversubscribed (Figure 3-6). Pursuit of many of the high-priority science goals discussed in Chapter 2 will be limited by both the current capabilities and the current capacity of NDSF assets. Continued expansion and diversification of the pool of potential users can be accommodated only by providing expanded access to needed assets.

4

Addressing the Need for Improved Deep Submergence Assets

HIGHLIGHTS

This chapter

- Explores options for providing better capabilities over greater geographic range (to support call for new ROV)
- Explores options for upgrading individual components (to support calls for a super-ROV and an enhanced HOV)
- Explores options for improving the overall capability of the asset pool (e.g., standard tool interfaces)

Research conducted in the ocean depths has important implications not only for understanding the oceans, but also for advancing some of the most basic and profound areas of human inquiry. In recognition of the significant potential this research holds and the unique and challenging requirements that work in the deep ocean represents, the National Science Foundation (NSF), National Oceanic and Atmospheric Administration (NOAA), and U.S. Navy have made a significant commitment to provide operational support for these efforts. Maximizing the return on the investment made in this unique and challenging scientific effort will require overcoming a number of natural and unique human obstacles. Yet, pursuit of many high-priority science goals will be limited by both the current capabilities and the current capacity of National Deep Submergence Facility (NDSF) assets. Continued expansion and diversification of the pool of potential users can be accommodated only by expanding access to needed assets. As is often the case, solutions to one set of problems may exacerbate others; thus, this chapter attempts to explore the various obstacles and their potential solutions in a holistic manner.

IMPROVED UTILIZATION OF EXISTING ASSETS

As discussed in Chapter 3 it appears that the scientific demand for deep submergence assets is, at present, not being adequately met. Part of this problem can be traced to an inadequacy in the number and capabilities of existing assets. Another part, however, can be attributed to the way in which existing assets are utilized. In particular, the current system does not always ensure a match between the requirements of federally funded projects and the appropriate deep submergence assets.

Modification of existing assets and construction of new assets to alleviate this problem would represent both a significant capital investment and an additional demand for operating funds. Decisions to commit these resources should be accompanied by a commitment to ensure the best use of the nation's deep submergence assets. The management of the nation's deep submergence assets should, therefore, be clarified and revised to ensure the optimal use of both existing and potential assets in future scientific research.

A Question of Access

Many previous studies have called for some reexamination of the use of U.S. and foreign platforms to support deep submergence science. However, detailed guidance beyond general calls for "increased access" have not been put forward. Thus, some additional discussion is warranted here.

As previously discussed, NDSF currently operates two vehicles: the HOV (human-occupied vehicle) *Alvin* and the ROV (remotely operated vehicle) *Jason II*. Support for the operation of these vehicles is guaranteed by the three NDSF sponsoring agencies (NSF, NOAA, and the Navy). The agencies are also major supporters of research that utilizes NDSF assets. For example, in 2002, NSF and NOAA accounted for nearly 100 percent of the operational days of all NDSF platforms, with NSF generally accounting for more than 70 to 80 percent annually. This arrangement ensures consistent support for NDSF, while allowing some flexibility in the burden borne by each of the three sponsoring agencies. This, in turn, allows each agency to predict its annual contribution to

NDSF funding and to develop budget mechanisms to accommodate that contribution.

The use of NDSF assets and University-National Oceanographic Laboratory System (UNOLS) vessels as surface support ships by scientists funded through NSF is covered by the funds from the Marine Operations Section of the NSF Division of Ocean Sciences (OCE). Proposals to NSF requesting the use of these vehicles are submitted to a number of science programs within OCE. No costs for the use of NDSF or other UNOLS platforms are included in the budgets of these individual proposals. This policy helps the evaluation process focus on the intrinsic scientific merit of proposals that may require deep submergence assets; it also ensures that assets that are already funded are utilized to the fullest extent possible. For the most part, vehicle use is essentially guaranteed to any project that is funded by the science program to which it was submitted. Scheduling vehicle use is facilitated in at least two ways. First, projects may be postponed to accommodate vehicle schedules. Second, one vehicle may be substituted for the other in projects where such substitution is possible. In some cases, it is simply not possible to accommodate vehicle requirements, and proposals may be rejected or deferred on these grounds.

Conversely, proposals to NSF that request the use of non-NDSF platforms must include platform operating costs in the proposal budget. Should the proposal be funded, funding for platform use must be provided by the science program to which the proposal was submitted. Since the additional cost for the use of these platforms (not including the cost of support ships) can be substantial ($10,000 to $30,000 per day of use), this additional cost is widely perceived as placing such proposals at an unfair disadvantage compared to those requesting NDSF platforms. Even when NDSF assets are available, a suitable UNOLS support ship may not be. In such instances the costs on non-UNOLS vessels may represent as much as one-third of the cost of project (RIDGE 2000, 2003). Under the current system, this additional cost must again be borne by the science program. The additional cost to the scientific programs of using non-NDSF or other non-UNOLS platforms is perceived by the scientific community as placing proposals calling for their use at a significant and unfair disadvantage. To the degree that this perception discourages scientists from more fully developing high-quality research programs that cannot be supported by the existing pool of NDSF assets, this lack of access to suitable assets is limiting the scope of deep submergence science (UNOLS, 1999). Steps should be taken to eliminate this effect if a meaningful expansion of deep submergence science is to be realized. Given that any action to develop additional NDSF assets will take at least two years, mechanisms for allowing limited access to non-NDSF (and other non-UNOLS) platforms should be explored.

Understanding the NDSF Cost Structure

Because the NDSF is funded irrespective of vehicle use, the marginal cost (i.e., the cost of an additional day of operation) is zero.[1] In contrast, the marginal cost of the use of non-NDSF assets can be substantial. From a fiscal perspective it is therefore sensible to require, when possible, that NDSF assets be used in favor of non-NDSF assets. In the absence of additional funds, excess demand for NDSF assets can be managed by a combination of asset substitution, scheduling, and if necessary, proposal rejection. If additional funds are available, excess demand can also be addressed by leasing non-NDSF assets.

There appear to be situations, however, in which deep submergence scientific goals cannot be met by NDSF assets but can be met by non-NDSF assets. Moreover, these limitations preferentially restrict the use of NDSF assets in certain areas of deep submergence science. For example, limitations on the viewing capability of *Alvin* and on its ability to achieve neutral buoyancy at multiple depths during a single dive make it less suited for certain types of midwater research than certain non-NDSF assets. In addition, the strong pressure to revisit specific sites on an annual or biannual basis, termed the "yo-yo" effect because *Atlantis* and *Alvin* have been pulled back and forth through the Panama Canal to and from these sites, limits access to *Alvin* in other geographical locations. For this reason, the valid fiscal argument favoring the use of NDSF assets has the unintended consequence of restricting the scope of deep submergence science.

In theory, one possible solution to this problem is to abolish the NDSF, sell its assets, and essentially create a free market for deep submergence scientific support. This solution, however, fails to capture the true economies of scale in the operation of deep submergence assets. On the opposite extreme, a solution would be to expand NDSF to cover funding for any deep submergence asset. This approach could reasonably be expected to increase overall costs dramatically, because all assets would become fixed-cost assets regardless of demand. A more reasonable solution that expands the scope of deep submergence science while capturing economies of scale is to upgrade the capabilities of the NDSF assets so that they can be used in all areas of deep submergence science. Although this is clearly an important part of the solution, by itself it may be inadequate. First, these upgrades, if they occur, will not be completed for two to three years, and some short-term measures are needed. Perhaps more impor-

---

[1]This, of course, assumes that all crew are salaried (no overtime) and there are no expendables. Even for instances in which this assumption is invalid, the marginal cost is still minimal when compared to the additional cost (beyond the fixed cost of maintaining NDSF equipment) of leasing equipment outside.

tantly, there is a danger that the existing pattern of use of NDSF assets will simply persist. One way to address both of these problems is to provide modest but immediate funding to support the use of non-NDSF assets. This funding might, but need not, reside within the NDSF, but to avoid disparate treatment, it should not be included in the budgets of the NSF/OCE scientific program. An additional benefit of establishing such a fund is that it would serve as a gauge of demand for capabilities not provided by current NDSF assets. NSF/OCE should establish a small pool of funds (on the order of 10 percent of the annual budget for NDSF) that could be targeted specifically to support the use of non-NDSF vehicles for high-quality, funded research when legitimate barriers to the use of NDSF assets can be demonstrated. If, as additional assets become available, the demand for non-NDSF vehicles declines (or never materializes), these funds could be used to address other (non-deep submergence) marine operational needs as determined by NSF/OCE.

DEVELOPING NEW ASSETS

Improved utilization of existing assets, although an important step, cannot be expected to fully address the scientific demand. As is apparent from Chapters 2 and 3, there is a need for different platforms to carry out different missions. Given the range of capabilities needed, this study concludes that the addition or enhancement to two specific platforms would most benefit science: a deepwater ROV system and an improved HOV. These platforms offer the most science for the current technology and, when used in conjunction with other platforms such as deep towed arrays, autonomous underwater vehicles (AUVs), ROVs, and other submersibles, provide tremendous wide-scale and localized-scale efficiency for performing scientific research. The addition of these assets would provide the greatest opportunity to meet the demands of world-class deep submergence science for scientific endeavors in all regions of the world and in more than 98 percent of the ocean's volume.

Construction of a New (7,000-m) ROV Platform

The present pattern of use (see Table 3-1) of existing scientific ROVs suggests that the greater geographic flexibility for ROVs comes when they can be operated in a "fly away" manner. Fly-away refers to an ROV that is not tied to a specific mother ship and can be relocated to a suitable ship of opportunity, making it an attractive tool for scientific research in various and diverse areas of the globe. As the portfolio of deep submergence studies diversifies in response to greater demand to conduct funded research in various regions of the world, additional demand for an already fully subscribed *Jason II* can be expected. As discussed in Chapter 3, NSF's Ocean Observatories Initiative (OOI) will provide a significant new capability to support deep ocean science and can be expected to have a significant impact on UNOLS fleet and deep submergence assets (NRC, 2003a). Although the placement of fixed observatories will relieve some demand for NDSF assets to support time-series studies, it does not appear likely that *Jason II* or *Alvin* will be undersubscribed in the near future. Furthermore, the demands for ROVs to support the construction and maintenance of the OOI network and the scientific programs associated with it will be significant. Assets specifically dedicated to support observatory science cannot be assumed to be available to also support expeditionary science. The NRC (2003a, p. 8) report concluded:

> *Without a commitment from NSF to augment ship and ROV capabilities to meet these needs, the scope and success of the ocean observatory program could be jeopardized and other types of ocean research requiring these assets could be negatively affected. UNOLS and its Deep Submergence Science Committee (DESSC) should develop a strategic plan that identifies the most cost-effective options for supplying the required ship and ROV assets for observatory operation and maintenance and NSF should commit the necessary funds to acquire these assets.*

Recommendations made in this report are exclusive of and in addition to those made for support of the OOI.

Characteristics of an Improved ROV System

Several factors should be considered when developing a new ROV system to support expeditionary science. Probably most important overall is that the new asset be considered as an overall system rather than simply as a new ROV. Such an approach supports the incorporation of several attributes that will greatly enhance its utility as well as its ability to complement existing assets. Some of the most significant attributes include standardized tooling suites, open software and hardware architectures, electric thruster systems, a variable ballast system, tether management systems, improved handling systems, and camera and lighting systems.

Standardized Tooling Suite. Interchange of tools, sensors, and equipment will be greatly enhanced by adopting standard interfaces between platforms. International standards such as those of the International Organization for Standards, American Petroleum Institute, and others should be agreed upon by the community and systems designed (or retrofitted) with standardized interfaces as a design parameter. Standard tools should be built to accommodate interchangeability and use with both ROV and HOV platforms. It is recognized that there will be special tools for specific one-time science missions, but standardization of interfaces will save design time, setup time, and operations and hence allow more time and resources to be devoted to research.

Open Software and Hardware Architecture. Open software architectures that incorporate object-oriented application modules will support faster and cheaper development, module reuse, shorter validation and verification efforts, and reduced maintenance costs over the life of the program. Use of commercial off-the-shelf (COTS) software for major infrastructure items (e.g., operating system, networking protocol) will let specialized onboard and offboard software systems take advantage of significant COTS software investments and advances. Open hardware architectures that follow commercial standards will likewise benefit from COTS product improvements while minimizing the impact of vendor volatility as long as attention is paid to the development of an adequate "middleware" layer that serves to isolate software applications from hardware components. Design and deployment of open architectures will require significant effort, but the benefits extend beyond single platform development. Benefits may extend not only across vehicles, but also across classes of vehicles (i.e., HOVs, ROVs, AUVs) and locales (i.e., both on the vehicle and at the mission management stations).

Greater Horsepower. This would be required to make the ROV more capable at depth for station keeping and moving into position. Providing greater horsepower to the ROV allows for more powerful thrusters for overcoming current and drag from the tether or umbilical cables as well as additional capacity for running tools, sensors, lights, hydraulic power units, video, and so on.

Electric Thruster System. An electric thruster system allows the ROV to be unobtrusive during operation. The Monterey Bay Aquarium Research Institute's (MBARI's) *Tiburon* ROV, which has all-electric thrusters with a hydraulic power pack for the operation of the arms and tooling skids, is an excellent example of this technology. Existing commercial electric systems are also available (e.g., the *Quest* ROV system from Alstom). These systems excel throughout the water column, whereas hydraulic systems tend to overheat at shallower depths.

Variable Ballast Systems. To perform midwater research and delicate studies just above a mudline on the seafloor and in boundary layers, a variable ballast system would be required. Such a system ensures greater visibility and more efficient use of dive time by allowing the submersible to trim at specific depths without the use of thrusters. This minimizes sediment and flow field disturbances and enables fine manipulation of fragile equipment, samples, and experiments even along a rough seafloor. The ability to maintain a stable altitude in the water column without thrusters and adjusting the speed of forward movement relative to suspended particles allow sinking particles to be followed through the water column. Sedimentation rates and study of the changes in chemical microenvironments around and within such particles can thus be measured along with sampling of the particles. The ability to float within any given water mass, moving in exactly the same speed and direction as the water mass itself, also gives "stealth" capabilities in the bioluminescent minefield that comprises the midwater environment. This is imperative for studies of bioluminescence and the effect of light on the ecology of the midwater realm. Measurements relating to fine-scale physical processes such as mixing vortices and interface structure and dynamics can be done from a stable platform through advanced dye injection techniques and accurate appraisals of the fine-scale distribution of plankton and other particles in the water column or benthic boundary layer. The mapping of vertically stratified fine-scale structures in the water column along a horizontal plane and measurements of associated chemical and biological variables can also be done only when a variable ballast system is employed.

Tether Management System (TMS). This hardware augments the variable ballast system and allows the system to be relocated to different ships, thereby alleviating the need for heave compensation of the power cable. A TMS with extended (long) tethers allows the ROV to "decouple" from the heavy and stiff armored umbilical and attenuates the effect of surface vessel induced reactions. A long (500 to 1,000m) neutrally buoyant, small-diameter tether reduces the drag on the ROV, allowing excursions further from the "footprint" of the surface vessel, and reduces the influence of the TMS for midwater- or seafloor-based work. This allows the ROV to reach large survey or exploration areas without repositioning the surface support vessel or requiring dynamic positioning capabilities. Docking the ROV to the TMS during launch and recovery allows for safer operations and expands the sea state window for launch and recovery.

Improved Handling Systems. Launch and recovery of the ROV are two of the most critical aspects of operations. Careful design of the handling systems will enable safer operations by reducing the dangers to personnel and equipment while providing the ability to operate in higher sea states. The system should be designed to deploy the ROV overboard with a minimum number of personnel. It should allow for "latching" the ROV-TMS to the launch system during launch and recovery operations to reduce movement of the ROV-TMS. When operating in harsh environmental conditions, consideration should be given to a "cursor" system on the support vessel. A cursor system is a rail-type system that allows the ROV to be deployed in heavy seas in a controlled manner. The ROV is controlled and guided to a position below the sea surface where it may then be deployed with less interference from the motion of the sea.

Improved Camera and Lighting Systems. Visualization of the ROV's environment is fundamental. High-definition (HD) cameras and associated lighting systems are essential pieces of equipment. A wide-angle or side-lighting illumination and highly light-sensitive cameras are needed for viewing the fine internal structures of gelatinous animals and large particle aggregates. The use of HD cameras will provide operators, observers, and researchers, via onboard and real-time satellite transmission, with greater definition of objects and the areas being viewed. Documentation will be enhanced, and imagery suitable for education and public outreach will also benefit. Multiple camera systems with overlapping fields of view to give real-time panoramic displays of the water column or ocean floor should also be incorporated to increase search efficiency and imaging capabilities.

Multiple-Frequency Sonar for Biological Studies. Recent developments in acoustic identification of marine life suggest that consideration be given to incorporating more sophisticated sonar systems into ROVs (and HOVs) intended to support midwater column work. Acoustic backscatter normally removed during signal processing of data collected for seafloor mapping can actually yield insights into the distribution of marine biomass in the water column. Furthermore, a fairly large database of species-specific acoustic signatures is being developed for many finfish species. Such techniques could greatly enhance the capabilities of ROVs conducting biological transects or other midwater column biological research. These systems may also have utility for mapping both diffuse and plume effluent from seeps and hydrothermal vents.

Use of Seven-function Arms. A seven-function spatially correspondent (SC) manipulator plus a "grabber" manipulator allows experienced as well as inexperienced scientists to control experiments that require a delicate touch. The use of seven-function SC manipulators (the commercial industry standard) will increase the ability of operators to perform precise tasks with a minimum of effort and in less time. The use of a second heavy grabber manipulator for placing large science packages, taking large samples, or moving objects should be considered.

External Data Links. The ability to "plug and play" new tools and to replace components quickly without breaking a pressure boundary allows external experimental apparatus to be plugged into the ROV electronics system without opening up the main electronics can. This is greatly facilitated by the use of an external data link housing and saves hours in set up and de-mobilize time.

Improved Reliability. Reliability of commercial ROVs has improved dramatically over the years. Some of the design and operating philosophy used by the commercial industry should be considered in any new design or refurbishment of science ROVs. For example, the use wherever possible of standard "off-the-shelf" equipment such as lights, cameras, manipulators, thrusters, winches, hydraulic actuators, connectors, computers, and instruments will increase the reliability of the ROV system and facilitate procurement of spare parts. Training of operators and technicians is paramount for improved reliability, as are operating and maintenance manuals, regular maintenance schedules, and adequate onboard spare parts.

The recommended details for this 7,000m ROV platform come from a wide body of work presented to the committee, found in the literature, or currently in use on scientific and commercial systems. Most are readily available off the shelf as components. The total cost of the system would be approximately $5 million, and it could be built and ready for service within one year of authorization.

Using the current UNOLS model, this ROV system could be mobilized to the current fleet without any significant additions of hardware, and the operational requirements would be similar to those for the current *Jason II* crew. Construction of a new ROV would not only address the current excess demand for *Jason II* but also significantly enhance the geographic range over which deep submergence science can be carried out. Although this study did not have the time or resources to explore the cost and benefits of operating the NDSF asset pool as a distributed facility, serious consideration should be given to basing any new ROV at a second location. Although there may be additional costs involved in having a second facility, such an arrangement could offer greater flexibility in cruise scheduling while minimizing any transit time required for periodic overhaul and refit.

Developing a More Capable HOV

In 1999, the deep submergence scientific community responded to a questionnaire circulated by Woods Hole Oceanographic Institution (WHOI) on improved submersible capabilities. Based on these results,

WHOI developed some preliminary specifications for an improved HOV (R. Brown, Woods Hole Oceanographic Institution, Woods Hole, Mass., written communication, 2003):

- Ascent or descent in two hours to 6,000m
- Improved viewport arrangement
- Increased sphere volume
- 4000-pound (in water) external science payload
- Battery capacity increase of 30 percent
- Automated position keeping in all axes
- Enhanced controls and monitoring for single-pilot control
- Pan and tilt video for each observer

Based on the preliminary requirements developed by WHOI, and augmented by input received during the course of this study, it is clear that additional improvements should be pursued in the following subsystems: (1) development of a mercury-free trim system due to environmental concerns; (2) complementary electronics and tooling fixtures that match with new 7,000m ROVs—this is critical for a cross-platform continuity of experiments; (3) optional fiber-optic cable to allow expanded communication with the surface for certain applications (e.g., video feed to other scientists, navigation data, surface control of cameras and equipment); (4) target localization, fixation, and following by tracking eye movements for video cameras onboard; and (5) a robust variable ballast system.

Based on discussions with scientists primarily interested in the seafloor, WHOI explored three options for developing an enhanced HOV. Since these were used as a starting point for consideration during this study, they are discussed here. In their simplest form they include the following:

- **Option 1:** An improved 4,500m *Alvin* using *Seacliff* or components of *Seacliff*
- **Option 2:** A 6,000m *Alvin* using the Lokomo hull
- **Option 3:** A new HOV (referred to as *Alvin II* in the WHOI report)

A summary of these three options as described by WHOI is presented in Table 3-2. There are a host of variations on these three options with cost estimates ranging from less than $1 million for a minor modification to $13 million for a completely new *Alvin*. The Committee on Future Needs in Deep Submergence Science has reviewed these options and discussed additional variations that were not included in the WHOI reports (Table 4-1). The committee also considered the feasibility of a full-ocean-depth HOV. It was determined that an 11,000m HOV was not a feasible option

TABLE 4-1 Comparison of HOV Vehicle Options

Features	Improved 4,500m *Alvin*	6,000m *Alvin* (Lokomo Hull)	New HOV
6,000m	No	Yes	Yes
Sphere			
Window arrangement	No change	Improved	Improved
Interior volume	No change	Increase	Increase
Comfort	No change	Increase	Increase
Science			
Electrical capacity	No change	No change	Increase
Payload	Increase	No change	Increase
Seafloor footprint	No change	No change	No change
Maneuvering Characteristics			
Ascent and descent rates	No change	?	Increase
Maneuverability	No change	No change	Increase
Speed	No change	No change	Increase (?)
Launch and recovery hazard	No change	No change	No change
Construction			
Maintainability	Increase	No change	Increase
Operating costs	No change	No change	No change
Variable ballast	Water	Water	Water
Trim	Hg	Hg	Water (?)
Ascent and Descent weights	Steel	Steel	Water (?)
Ship Infrastructure			
HOV weight	Decrease (?)	Increase	No change (?)
A-frame mods	No	?	No
Hanger usable	Yes	Yes	Yes
Cost	$686,800	$4,207,405	$13,248,670

SOURCE: Modified from Brown et al., 2000.

because design and construction of such a vessel could not be completed within the two-to three-year time frame or within the cost constraints provided by NSF (see the statement of task). Specifically, for an equal-size sphere, nearly doubling the depth capability from 6,500 to 11,000m, would almost double the pressure hull weight. For the new 6,500m HOV the titanium hull weight is approximately 11,000 pounds, or one-third the total HOV weight of 32,000 pounds. Nearly doubling the pressure sphere weight would almost double the full-ocean-depth HOV weight and place it well beyond the capacity of *Atlantis*. Modifying *Atlantis* would place the total cost well beyond the NSF budget.

In addition, it is not clear if syntactic foam, batteries, or electronic components can be built and certified for human occupancy for such depths. Even if all components were available off the shelf there is no certification test facility for such pressures. For these reasons, consideration was not given to construction of a full-ocean-depth HOV. Rather, attention was focused on evaluating and supplementing (as needed) options presented initially by WHOI (Brown et al., 2000). The three options[2] are summarized and assessed in the following material.

Option 1. Refitting the *Seacliff*

Use of *Seacliff* is deemed to not be feasible for a number of reasons. First, the U.S. Navy has indicated that it does not want *Seacliff* modified. Second, even if it were available to be modified, the tending vessel, *Atlantis*, can only support a submersible of 40,000 pounds or less (in air). The cost of modification of *Atlantis* to accommodate the 66,000-pound (in air) *Seacliff* would be prohibitive. Third, *Seacliff* was designed based on the old *Alvin*. The only new capability that *Seacliff* provides beyond the existing *Alvin* is increased depth capability from 4,500 to 6,000m. Hence, *Seacliff* suffers from most of the same limitations, which would be reduced or eliminated in the new *Alvin* design. These limitations include descent time, viewport arrangement, smaller sphere volume, payload, battery capacity, and position keeping. Fourth, *Seacliff* is significantly larger in size than *Alvin* and cannot maneuver in some of the tight locations that are needed for exploration of newly created seafloors. Because of these considerations, the WHOI study focused on either fabrication of a new *Alvin* or modification of the existing *Alvin*.

Option 2. Modification of the Existing *Alvin*

The WHOI cost estimates of modifications to *Alvin* vary from $700,000 to $4.2 million, depending on the extent of modification. However, according to the WHOI assessment, none of the existing *Alvin* modifications improve on the electrical capacity, the ascent-descent rates, the maneuverability, the speed, or the launch-recovery hazard. In addition, modifying the existing *Alvin* will not provide all of the capabilities needed to

[2]The term "option" is used when discussing the WHOI report (Brown et al., 2000). The term "approaches" is used to designate scenarios explored during the current study.

undertake some of the high-priority deep submergence science discussed in Chapter 2. Finally, it must be recognized that modifying the *Alvin* sphere will take the vehicle off-line for some time and put it at risk for the long term should technical issues arise in sphere modifications.

Option 3. Construction of a New *Alvin*

WHOI's estimated cost for a new *Alvin* weighing 32,000 pounds is $13.25 million or just over $400 per pound, compared with the *Jason II* at $320 per pound (Appendix D). Thus, the estimated cost of a new *Alvin* is judged to be realistic. The higher per-pound cost compared with *Jason II* is due to the engineering and testing of a human-certified system (e.g., life support, power supply, pressure hull) of the new *Alvin* compared an ROV.

Since the $13.25 million cost is based on estimates of the costs of the individual subsystems of a new HOV, the total cost with design, assembly, certification, contingencies, and initial sea trials would likely be in the $15 million to $16 million range. Thus, a new HOV can be expected to be built within the NSF budget limitations.

APPROACHES TO CONSTRUCTING A NEW HOV: THE CURRENT STUDY

Based on additional input obtained during this study, the following additional desirable features for a new DSV should be incorporated:

- Upgrade of variable ballast system. This is a high-priority need if the new HOV is to perform midwater research. Current systems on *Alvin* do not allow for trim-out repeatedly at multiple depths within a 1,000m envelope.
- Development of a mercury-free trim system for environmental reasons.
- Complementary electronics and tooling fixtures that match with a new 7,000m ROV. This is critical for a cross-platform continuity of experiments.
- Fiber-optic cable. This allows continuous data streaming for the HOV (experiments, video feed to other scientists, navigation data, etc.) The line can also be used to control cameras and equipment from topside.
- Target localization, fixation, and following by tracking eye movements for video cameras onboard.

Design of an HOV begins with the pressure hull since the weight and balance of the remainder the vessel are controlled by the sphere, which is approximately one-third of the total weight of the HOV. In terms of maintenance, titanium is the material of choice since it is highly corrosion resistant in seawater. The existing *Alvin* sphere is titanium alloy 6211, a special naval alloy that has not been produced for 20 years. Very high strength maraging steels such as those used on the *Mir* HOVs give strength and density performance similar to titanium, but they are susceptible to corrosion and must be maintained carefully to prevent hydrogen embrittlement.

In constructing a new *Alvin*, there are four approaches to consider:

1. A new 6,500m titanium pressure sphere
2. Purchase of the existing Lokomo steel pressure sphere
3. Modification of the existing *Alvin* titanium sphere with improved viewports
4. Use of the existing *Alvin* pressure sphere as is.

Approaches 1 and 2 would provide increased depth capability of a new HOV. In addition, these approaches for building an entirely new HOV provide a course that would leave the *Alvin* intact and in reserve if funding for a second support ship should become available. This is currently not an option for NSF but could be in the future. Refitting the *Alvin* could be less costly than building an entirely new second submersible 10 to 20 years into the future. Approaches 3 and 4 would limit the *Alvin* depth capability to 4,500m. These four approaches are presented in order of both decreasing overall scientific capability and decreasing design risk. In other words, while each successive approach will provide an HOV with less overall capability, it also represents an approach with fewer engineering hurdles to overcome, suggesting that it is more likely to be achieved in a predictable amount of time and for a predictable amount of funds. For example, Approach 4 calls for using the *Alvin* sphere with a complete rebuilding or replacement of the external components (e.g., syntactic foam, variable ballast system, thrusters). This would provide a less capable HOV, in terms of both its depth capability and the viewport arrangement. Given that the *Alvin* is Navy property and that modification of the subsystems would not be difficult, this is the most predictable path to a new HOV.

Increased depth capability permits operation over a greater portion of the ocean. Equally important, enhanced depth capability permits midwater work in these same regions due to the safety requirement that the *Alvin* not operate (even at midwater) in any location in which the bottom is greater than the depth capacity of the pressure sphere. In addition, greater depth capacity means higher overall performance at all depths because the required battery capacity, payload, and ascent-descent rates, for example, needed for operating at greater depths (greater than 4,500m) equal improved performance time at shallower depths.

The greatest risk in meeting the cost estimate of a new HOV is the cost of fabricating the pressure hull, which has been estimated by WHOI at $2 million out of the 2001 estimate of $13 million. The sources of the original titanium 6211 plate and the forging facilities used to fabricate the existing *Alvin* and *Seacliff* titanium pressure hulls in the 1970s no longer exist. The United States has extremely limited industrial experience in welding heavy section titanium. The former Soviet Union and Japan have extensive capabilities for such welding, and this option should be considered. Although WHOI has obtained several estimates for forging new hemispheres and some existing titanium 6211 plate may be available, the source of the plate is not firm and the forging and the welding vendors do not have prior experience with welding a pressure hull of heavy section titanium. A failure during welding could result in loss of the entire sphere and a $2 million increased cost of obtaining a new pressure hull. NSF/OCE should plan carefully for such a contingency.

Approach 1: A New Titanium Sphere

Titanium has much lower long-term maintenance costs than steel. The original *Alvin* sphere was built of steel in 1964 but upgraded to titanium in 1973. If fabrication facilities for a new titanium sphere can be confirmed at a reasonable price, a larger-diameter, greater-depth-capacity, better-viewport-arrangement titanium sphere is the most desirable option for the new HOV because it will provide the greatest science capability.

Nevertheless, there are significant risks to production of a new titanium sphere. As noted above, there is no industrial experience in the United States for acquiring the plate, forging it into hemispheres, and welding the hemispheres together. While WHOI has been successful in obtaining cost estimates for each of these fabrication steps, it is not clear that these cost estimates will be maintained when a formal price quotation is requested. Experience with other large-scale science projects with unique requirements has shown a reluctance on the part of upper-level industry management to accept the risks of one-of-a-kind fabrication when compared to the potential profits. In addition, the two welding options have a significant cost risk. It required a year to weld the *Seacliff* using gas tungsten arc welding (GTAW) at Mare Island Naval Shipyard. Should it take on the order of a year to weld the new sphere, the welding cost estimates will likely be exceeded. The seemingly preferred option of electron beam welding, although much faster, has the potential to produce a nonreparable joint if anything goes wrong during the hour-long weld process. For this reason, NSF/OCE must select and fabricate the pressure hull before final design and construction of the remainder of the new HOV proceeds too far. The relative risk of acquiring a new titanium pressure sphere is judged to be high.

Approach 2: Use of the Lokomo Steel Sphere

If the new titanium sphere is not judged practical due to either excessive costs or failure to confirm fabrication contracts, the existing steel sphere at Lokomo in Finland should be considered. This sphere provides the enhanced depth capacity and the larger sphere volume desired in a new HOV. It should be remembered that Lokomo built four spheres (using casting technology). One was scrapped and two are used in the *Mir* HOVs. It can be assumed that the third sphere was the third-highest quality of the acceptable spheres. In considering use of the Lokomo sphere, NSF must carefully weigh the cost of the sphere compared to the higher long-term maintenance costs. The relative risk of the Lokomo sphere meeting all quality requirements is considered moderate.

Approach 3: Modification of the Existing *Alvin* Pressure Hull

It is probable that the existing *Alvin* sphere can be modified with new viewports. Such an option does not enhance depth capacity or sphere volume, but it is more likely that a realistic price quotation can be obtained for this smaller modification as compared with a completely new titanium sphere (Approach 1). It is most likely that GTAW would be used to install the new viewports. Such welding could likely be completed within six months. Electron beam welding would be an option as well and could be repaired in this case by reforging a larger insert sector if the first weld is deficient. Such is not the case with welding two hemispheres (Approach 1), where cutting out a defective weld results in two inadequately sized hemispheres. The relative risk of modifying the existing *Alvin* sphere is considered moderate. However, if Approaches 1 and 2 are not deemed feasible, a failure in fabricating Approach 3 would leave NSF without any pressure hull.

Approach 4: Use of the Existing *Alvin* Sphere As Is

There is little risk in reusing the existing *Alvin* sphere and rebuilding all of the internal and external components surrounding it. The costs of this option should be well within the NSF budget capability. The sacrifices of this option are depth capability, sphere volume, and viewport arrangement. This fourth approach for a new HOV will provide all of the desired improvements, except these three, and the risk is low. Thus, it is believed that modification of the existing *Alvin* (Option 2) would not provide the best science capability for the cost, compared to a new HOV that uses the existing *Alvin* sphere.

In the long run, fabricating a new titanium sphere may be high risk, but represents less overall risk since the existing *Alvin* sphere would not be destroyed. In addition, the deep submergence engineering community would have obtained valuable experience in fabricating the new titanium sphere, whether it is successful or not. The most promising approaches for moving ahead during the time frame articulated by NSF/OCE would make use of one of the two existing spheres. The first is the third, unused sphere from the Russian *Mir* HOV series (referred to as the Lokomo sphere), which has been rated to 6,000m and has a better configuration of portholes allowing both an expanded field of view and the maximum diameter of the sphere to be used for human occupancy—thereby somewhat relieving the cramped conditions within the sphere. Given the lack of a clear and overwhelming scientific argument for conducting HOV operations at depths greater than 4,500m, it is not obvious that significant resources (in excess of those needed to fully upgrade the current *Alvin*) should be expended (i.e., although a deeper-diving HOV would be considered highly desirable, it is not clear that it would be essential). However, viewing and internal space configurations would benefit. Thus, constructing an HOV capable of operating at significantly greater depths (6,000m plus) should be undertaken only if additional design studies demonstrate that this capability can be delivered for a relatively small increase in cost and risk.

Innovative Technological Advances

Given a major investment in new assets, innovative technological advances should be incorporated into new HOVs, ROVs, and AUVs whenever practical to support current and future research needs of the scientific community. Enhancements (e.g., better cameras, lights, communications, computational platforms, tool pools) should be made to expand the platforms capabilities and useful lifetime, reducing the pressure to build new systems in the near future. A number of promising technologies for future ocean science needs have been identified based on recommendations in the following reports: UNOLS (1994, 1999) and NRC (1996, 2003a). They are discussed later, but first these reports are incorporated into the overall context of mission management since this drives the basic functional requirements, which in turn drive technology development and application efforts.

Mission Management

Much of HOV mission management is done manually once the initial pre-mission planning is completed. ROVs and AUVs, however, can be operated along a spectrum, ranging from manual *tele-operation*, to moderate levels of *automation*, to high levels of *autonomy*. Tele-operation places the operator "in" the loop, relying on onboard sensors to assist in positioning, rather than viewing the vehicle's position directly via a viewport. With moderate levels of automation (e.g., an autopilot), the required control bandwidth is reduced and the operator is less intensively engaged in *inner loop control*. With higher levels of automation (e.g., waypoint navigation open-loop raster scan area search behaviors) operator engagement is sporadic—monitoring overall mission progress and setting high-level platform goals or behaviors in a supervisory control mode of interaction (Sheridan, 2002). Being "in" the loop, the operator can step in when or if the onboard levels of automation cannot deal with a wide range of contingencies (e.g., an unanticipated subsystem failure) to override the vehicle automation and/or reconfigure the system to ensure mission success or, at least, vehicle recovery. Dealing with a broad range of contingencies—unanticipated and/or without "response scripts"—requires a level of onboard intelligent behavior that is labeled *autonomy*: situational awareness, engagement in (some) reasoning under uncertainty, and decision making (to generate plans) with extremely limited or no operator intervention. Platform autonomy puts the operator effectively "out of" the loop.

Technological advances, some of which are described below, establish the level of automation along a spectrum, but of equal importance are issues concerning the manner in which the operator interacts with the system, independent of the level of autonomy. As identified in a recent U.S. Air Force study on unoccupied aerial vehicles, this area of human-system integration is not merely about the operator controls and displays (the human-computer interface, or HCI), but also about the functional interactions between the human and the system performing the operations (U.S. Air Force Scientific Advisory Board, 2003). At low levels of automation (high levels of tele-operation), the *display demands* become paramount: maintaining operator awareness — particularly spatial awareness—requires high levels of sensory fidelity (e.g., resolution, field of view, depth of view, illumination, color), with high demands on onboard sensors, transmission bandwidth and latency, and displays made available to the operator. These effectively "max out" when demands are made for virtual tele-presence; that is, a perceptual sensation indistinguishable from direct observation. *Control demands* are probably maximal as well since fast control loops have to be maintained, but these are minimal demands when compared with the imaging sensor and download requirements. *Functional demands* are all minimized in terms of what the vehicle has to provide, since the operator is doing most of the work. This, of course, results in maximum operator workload. With high levels of platform autonomy at the opposite end of the spectrum, operator *display or control demands* (and communications requirements) are minimal. *Functional demands* for the operator are low, but this comes at the expense of extremely sophisticated software architectures and algorithms underlying "intelligent vehicle behaviors." On the basis of the previously cited Air Force study, much of this functionality is still years away, even in the R&D arena. In the middle of the spectrum is the potential for providing maximum flexibility across all three areas: displays, controls, and functionality. This calls for an HCI that lets the operator access the loop as needed, across activities that range, for example, from intermittent supervisory control of vehicle operations to intensive tele-operation of a platform effector via task-configurable displays and controls. *Functional capabilities* also must be flexible to support a variable "locus of control" between the operator and the platform. Adaptive function allocation is one approach to this problem.

Enabling Technologies

Significant gains could be made in underwater operations regardless of the class of platform (HOV, ROV, or AUV) if more attention were paid to upgrading non-platform-specific assets, particularly in the areas shown in Table 4-2.

In the platform payload area, gains are possible by developing or adapting *sensors* that are smaller, use less power, and are modular. Low-

TABLE 4-2 Enabling Technology Areas

General Area	Specific Need
Platform payload	Sensors (especially optical imaging sensors) Actuators, tools, and effectors Onboard processors and associated software architectures and algorithms
Communications	Between platform and surface Interplatform links
Operator display or control stations for mission management	Displays and controls Decision aiding and automation

power miniaturized sensors would particularly benefit HOVs and AUVs with their limited onboard power budgets, but would also reduce tether power requirements for ROVs. Modular sensors would benefit all three classes of platforms because development costs are spread over a larger population of customers and reliability would be expected to increase over time with a larger user base. This would necessitate a greater emphasis on standardization regarding mechanical, electrical, and software interfaces, but the upfront costs would be rapidly recovered in recurring development and operating savings. A similar argument can be made for standardization of interfaces for actuators, tools, and effectors.

Because of the critical importance of optical imaging sensors, they are given separate treatment in an expanded section at the end of this chapter. An area that shows particular promise for advancement is that of *onboard processing and associated software architectures and algorithms*. Platform-independent needs clearly exist in the following areas:

- Sensor processing, data fusion, machine-based perception, and situation awareness
- Vehicle health management, including failure detection, diagnosis, and recovery or reconfiguration
- Communications systems management to deal with lost link contingencies
- Mission planning, replanning for known contingencies, and dynamic replanning for opportunistic situations
- Improved operator display or control stations (for HOVs)
- Multivehicle coordination and cooperation algorithms and decision aids.

To support these functional areas, and reduce the cost or time of developing future systems, there exists a real need for *modular open software-hardware architectures*. The current trend for advanced embedded systems is to work within a layered architecture (see Table 4-3) that makes maximum use of COTS hardware and software, while allowing the development and use of mission- or domain-specialized application software modules or application packages.

In the realm of communications, problems and solutions are driven strongly by the presence or absence of a tether rather than the type of platform under consideration. A tethered HOV, utilizing a small-diameter fiber-optic cable to transmit video and data back to the ship, could have considerable advantages for data gathering and recording, observation (by allowing more eyes on target), access to topside information (e.g., digital maps), and processing power (e.g., offboard computational facilities). It is meant to be unobtrusive and to augment the mission and is

TABLE 4-3 Layered Software-Hardware Architecture

Layer	Function and Benefit
Application	Provides the functionality needed for the mission and the vehicle via modular object-oriented components. Proper design will support migration of components from one mission or vehicle to another, thus reducing code development and validation efforts
Application programmer interface (API) middleware	Isolates the application layer from the operating system below it, acting like a virtual machine for the application software
Operating system (OS)	Runs the API middleware and controls the hardware components of the computing platform(s). Use of COTS products (e.g., Linux, MS Windows) significantly reduces costs and enables rapid upgrading
Computing hardware	Runs the OS layer and consists of the main board, central processing unit (CPU), memory, input/output ports, and so on. Use of COTS products (e.g., Pentium CPU) significantly reduces costs and enables rapid upgrading
Platform networking layer	Provides connectivity between computing platforms, enabling distributed computing. Use of COTS products (e.g., TCP/IP [Transmission Control Protocol/Internet Protocol], Ethernet) significantly reduces costs and ensures interoperability

therefore an optional capability especially for dives below 3,000m, unless a bottom station is provided, as per the exploratory full-ocean-depth AUV-ROV hybrid proposal from NSDF. It is important to note that a small but nontrivial proportion of the dives may lose communication with the surface due to snapping of the single optic cable. It is a facultative hybrid system rather than obligate. Conversely, since untethered AUVs must rely on bandwidth-limited, fragile, and point-to-point acoustic links, they could clearly benefit from greater investments in onboard autonomy, allowing near-real-time decisions to be made locally, making low-bandwidth possible while retaining high-level operator control of the platform's operations. ROVs pose an intermediate load on the communications links: because they are tethered they can support high-bandwidth downloads of local imagery to the operator, as well as low-bandwidth control downloads from the operator. However, they also impose a high workload on the operator because of the necessity of keeping him or her "in the loop."

In the operator display and control station area, significant improvements can be made, both onboard and offboard. Improved displays can be developed by accelerating the evolution from dedicated displays where a given display is dedicated to one or a few variables, to integrated multifunction displays where a single piece of display hardware (e.g., a liquid-crystal display [LCD] panel) can support different display functions (e.g., video imagery, subsystem status). Both provide an integrated presentation of the related variables, typically via graphical means. There is a long history of this type of display evolution in the aerospace industry, and the HOV-ROV-AUV industry could benefit from the lessons learned there. There is currently disagreement among users from different fields (e.g., biological observers vs. geological) as to whether records of observations should even have date-time-heading-depth information burned onto the records. This can be avoided through the use of digital watermarks.

Improved controls could also be developed in a similar fashion via an evolution from dedicated switches and buttons to computer-based virtual analogs. Where physical controls are still deemed necessary (e.g., a hand controller for a remote manipulator), multifunction controls would not only save precious space (in an HOV) but also support more integrated operations by the pilot or operator. Again, the aerospace industry, particularly in the high-performance military cockpit, has been a pioneer in this area, and significant benefits could accrue from taking a more holistic approach to HCI workspace design. Finally, improved *decision aids and automation support* could enhance operations and reduce operator workload across a range of vehicles, especially legacy HOVs and ROVs. Some promising functional areas have been noted, in the discussion of onboard processing and software architectures or algorithms; these are just as valid for the offboard operator or supervisor. Naturally, the same arguments hold true here for the use of a layered software-hardware architecture, just as they did in the earlier discussion of on-platform requirements. Moreover, use of the *same* architecture for both onboard and offboard computing is strongly recommended, because it not only will lead to reduced development and validation costs overall, but also will support flexibility in assigning the "locus of control" (i.e., on-platform versus off-platform) as missions and technical capabilities change and evolve over time.

Optical Imaging Sensors

The strongest argument for HOVs is that there is no replacement for in situ human three-dimensional visualization and situation awareness. Human sensing and visual data processing allows extraction of directly relevant and needed task information. Human presence on the scene allows immediate understanding of the three-dimensional global aspects of a site and subsequent evaluation and assessment of relationships and interactions. Numerous scientific discoveries would not have been possible were it not for observations made by scientists aboard *Alvin* and other HOVs. Other advantages of HOVs include: requiring a shorter time to carry out the same exploration in a given ocean, generating excitement, providing educational experience, and breeding creativity and energetic scientific thinking. HOVs will continue, however, to have viewport limitations due to vehicle design constraints, and the associated restricted field of view (FOV) significantly reduces the advantages of human three-dimensional visualization. Furthermore, the heavy reliance on direct operator vision for HOV operations can lead to observations that are undocumented by conventional still or video imagery. Even when such imagery is recorded, the limited FOV of the cameras can provide instantaneous coverage over only a small section of the scene. To address these limitations, it is critical that a capability exists to generate high-resolution real-time panoramic imagery of the surrounding environment. Virtual and augmented reality systems and tele-presence are other evolving technologies that would provide tremendous potential for HOV as well as ROV operations.

Panoramic Imaging. Three existing imaging technologies provide a large field of view, ideally 360 degrees: (1) catadioptric sensors using a combination of lenses and mirrors in carefully arranged configurations relative to a standard camera (Nayar, 1997; Peleg et al., 2001); (2) collection of single-line scans or image strips from a rotating camera; (3) alignment of images from several cameras (Nalwa, 1996; Swaminathan and Nayar, 2000; Negahdaripour et al., 2001; Neumann et al., 2001; Firoozfam and Negahdaripour, 2002, 2003). A catadioptric sensor (e.g., commercially available omnidirectional design) can be deployed in a waterproof housing but has low resolution because the total pixels of a single camera are distributed over the panoramic view. High-resolution scanning systems can be either slow in imaging fast dynamical events or costly with the high-speed precise electromechanical scanning components required to reduce motion blur. Multicamera design—each camera covering a small section of the entire view—is an attractive solution for high-resolution precision imagery at modest cost. Multiple cameras are currently available on most vehicles but are employed primarily to switch between views. Generating panoramic views requires strategic placement of an array of CCD (charge-coupled device) cameras and sensors, synchronized image capture, and alignment of digitized images based on photo-mosaicing techniques. The alignment is perfect beyond a typically small "minimum working distance" (Swaminathan and Nayar, 2000). Such seamless panoramas can be constructed on the vehicle in real time with high-speed image acquisition systems and today's PC-type processors (currently used for tele-conferencing and video surveillance [Nalwa, 1996; McCutchen, 1997]). If the required hardware cannot be installed on the vehicle, raw video from various cameras can be transmitted via the tether and processed on the surface. Though not currently available off the shelf, enabling technology based on a single fiber-optic cable to transmit multiple high-definition quality video streams can be developed. With panoramic imaging technology in place, Moore's Law (the number of transistors per square inch of circuit will double every year) will ensure the ability to incorporate more and more complex operations to improve resolution and process images in real time for other capabilities (e.g., track moving fish). Finally, stereo images can be generated from a single camera with mirror reflections, enabling three-dimensional panoramic visualization at the rate of several frames per second (Gluckman and Nayar, 2000). This technology can be deployed underwater with proper housing and calibration, significantly enhancing ROV or HOV operations, as well as supporting various NSF initiatives for education and broad outreach.

Virtual reality. Virtual or augmented reality (VR) and tele-presence are other evolving technologies for enhanced three-dimensional views, peripheral vision, situation awareness, and hand-eye coordination for manipulation tasks and education in the submergence sciences. VR systems substitute real information received by human senses for artificially generated input, creating the impression of presence in a virtual environment. With a head-mounted display and computer generation of accurate sensory information, a user can be given the impression of being immersed in the virtual environment (immersive virtual reality) in order to navigate and manipulate objects in that world. VR systems are currently utilized in medicine, education and training, three-dimensional scientific visualization (biochemistry, engineering), computer-aided design, tele-operation and tele-manipulation, military applications (flight simulators), art (visual, musical), games, and entertainment. In non-immersive VR systems, the user views the virtual world through monitor(s) or projection screen(s). As the user moves, the view adjusts accordingly. "Augmented reality" is often sufficient, replacing the real world (not completely but only in certain aspects) with elements of the virtual environment. Realization of VR systems is supported by the combination of several technologies such as advanced (fast) computers, advanced computer communication networks, and HCIs. Fast computer networks enable exchange of information between remote locations, specifically, between the sensor locations and the operator displays.

Tele-presence (or virtual presence) systems extend the operator's sensory-motor facilities and problem-solving abilities to a remote environment. In tele-operation, the operator also has the ability to perform certain actions or manipulations at the remote site. Traditionally, these have required powerful general-purpose computers (e.g., Silicon Graphics, Suns) and graphical subsystems for real-time rendering and display of virtual environments that generate the visual stimulus. Recent developments in PC hardware and software have increased CPU (central processing unit), memory, and rendering capabilities considerably, allowing many legacy applications to migrate to much cheaper and more reliable platforms.

Distributed VR is the enabling technology for interactions within the same virtual world of geographically distributed networked users. Each networked computer tracks the actions of one user and creates the illusion of the user's presence in the shared virtual environment. Development of virtual or augmented reality systems for deep submergence science should be a high-priority research initiative that will address a number of inherent technical complexities including, but not limited to, visual artifacts from the movement of lights with cameras; the participating medium (water), which complicates the three-dimensional reconstruction; and the need for a very large optical dynamic range that approaches the dark-adaptation capability of the human eye to create realistic views.

CONCLUSION

Although current assets within NDSF are not adequate (in terms of quantity, capacity, and capability) to support some of the most promising areas of deep submergence science, several viable options exist. The most immediate need may be addressed by expanding access to non-NDSF assets by providing limited funds to support research efforts in instances where use of NDSF assets can be demonstrated to be a nonviable option. Given the limited capacity of the most capable non-NDSF assets, this option cannot be considered a satisfactory long-term solution. Thus, the construction of a more capable ROV system as well as a more capable HOV is recommended.

As these new assets are developed, significant effort should go into pursuing designs that incorporate many state-of-the-art capabilities. In short, while *Jason II* has proven itself to be an important and reliable component of the NDSF, designs for the new ROV systems called for here should incorporate a wider suite of capabilities than were incorporated in the design of *Jason II*. This report identifies a number of promising technologies, not all of which are mature enough to be incorporated into any ROV or HOV built in the next two to three years. Those technologies under development in marine technology, aerospace engineering, and military fields that are not yet mature enough for immediate implementation should be followed and considered for future generations of ROVs or HOVs.

5

Summary and Recommendations

HIGHLIGHTS

This chapter

- Articulates justification for and recommends improving access to, and the utilization of, the nation's deep submergence assets
- Articulates justification for and recommends construction of a new ROV
- Articulates justification for and recommends delivery of an upgraded HOV

As Chapter 2 points out, beyond its importance to ocean science, research in the deep ocean touches on some of the most basic scientific questions facing humanity. In recognition of both the importance of this work and the unique challenges that must be faced to pursue it, the nation has made a significant commitment to provide operational support. To maximize the scientific return on investment in this area it will be necessary to overcome both technological and institutional obstacles. This chapter discusses the main obstacles and potential solutions.

As shown in Chapter 4, the scientific demand for deep submergence assets is, at present, not being adequately met. Part of this problem can be traced to the inadequacy of the number and capabilities of existing assets to perform the type of scientific effort associated with deep submergence science funded through the National Science Foundation (NSF) and the National Oceanic and Atmospheric Administration (NOAA). This report makes specific suggestions for additions to and modifications of this asset pool. Part of the problem, however, can be attributed to the way in which existing assets are managed and utilized. In this context, management refers to the acquisition, maintenance, and accessibility of vehicles used to carry out basic ocean research and exploration. In particular, the current management system does not always ensure a match between the requirements of federally funded projects and the appropriate deep submergence assets. Modification of existing assets and construction of new assets to alleviate this problem would represent a significant capital investment and would likely engender additional demand for operating funds. Decisions to commit these resources should be accompanied by a commitment to ensure the best use of the nation's deep submergence assets. The management of the nation's deep submergence assets should, therefore, be clarified and revised to ensure the optimal use of both existing and potential assets in future scientific research. The following discussion points out key steps that should be taken to ensure that basic research conducted at depth in the ocean is not limited by access to appropriate research platforms.

PROBLEMS WITH ACCESSIBILITY

Many previous studies have called for a reexamination of the use of U.S. and foreign platforms to support deep submergence science. Beyond general calls for increased access, however, specific suggestions have not been put forward. Thus, some additional discussion is warranted here.

As discussed in Chapter 1, the National Deep Submergence Facility (NDSF) currently operates two vehicles: the human-occupied vehicle (HOV) *Alvin* and the remotely operated vehicle (ROV) *Jason II*. Support for the operation of these vehicles is guaranteed by the three NDSF sponsoring agencies (NSF, NOAA, and the U.S. Navy). These agencies are also major supporters of research that utilizes NDSF assets. For example, in 2002, NSF and NOAA accounted for nearly 100 percent of the operational days of all NDSF platforms, with NSF generally accounting for more than 70 to 80 percent annually. This arrangement ensures consistent support for NDSF, while allowing some flexibility in the burden borne by each of the three sponsoring agencies. This, in turn, allows each agency to predict its annual contribution to NDSF funding and to develop budget mechanisms to accommodate that contribution.

The use of NDSF assets by scientists funded through NSF is covered by funds from the Marine Operations Section of NSF's Ocean Sciences Division (OCE). Proposals to the NSF requesting use of these vehicles are submitted to a number of science programs within OCE. No costs for the use of NDSF platforms are included in the budgets of individual proposals. The cost of NDSF platform use is borne by the Marine Operations Section, rather than by individual science programs. This policy reflects the commitment made by NSF to provide the additional fiscal support needed to undertake deep ocean research and ensures that proposals are evaluated on intrinsic scientific merit. For the most part, vehicle use is essentially guaranteed to any project that is funded by the science program to which it was submitted. Scheduling vehicle use is facilitated in at least two ways. First, projects may be postponed to accommodate vehicle schedules. Second, one vehicle may be substituted for the other in projects where such substitution is possible. In some cases, it is simply not possible to accommodate vehicle requirements, and proposals are deferred or even rejected on these grounds.

Proposals to the NSF that request the use of non-NDSF platforms must include platform operating costs in the proposed budget. Should the proposal be funded, funding for platform use must be provided by the science program to which the proposal was submitted. Since the additional cost for the use of these platforms (not including cost of support ships) can be substantial ($10,000 to $30,000 per day of use), this compared cost is widely perceived as placing such proposals at an unfair disadvantage to those requesting NDSF platforms. This lack of access to suitable assets outside the NDSF is limiting the scope of deep submergence science (UNOLS, 1999). It is apparent that realizing the vision of deep ocean research described in Chapter 2 will require access to a broader mix of more capable vehicles than are currently available through the NDSF.[1]

Because the NDSF is funded irrespective of vehicle use, the marginal cost (i.e., cost of an additional day of operation) is zero. In contrast, the marginal cost of using non-NDSF assets can be substantial. From a fiscal perspective, it is therefore sensible to require, when possible, that NDSF assets be used in favor of non-NDSF assets. In the absence of additional funds, excess demand for NDSF assets can be managed by a combination of asset substitution (ROV for HOV or vice versa), scheduling, and if necessary, proposal rejection. If additional funds were to be made available, excess demand could also be addressed by leasing non-NDSF assets.

There appear to be situations, however, in which deep submergence scientific goals cannot be met by NDSF assets but can be met by non-

[1]The submersibles used to support deep ocean research are similar to those discussed in two recent reports *Enabling Ocean Research in the 21st Century: Implementation of a Network of Ocean Observatories* (NRC, 2003a) and *Exploration of the Seas: Voyage into the Unknown* (NRC, 2003b). The recommendations in this report are above and beyond any capabilities called for in those two reports.

NDSF assets. Moreover, these limitations preferentially restrict the utility of NDSF assets in certain areas of deep submergence science funded by NSF. For example, limitations on the viewing capability of *Alvin* and its capability to achieve neutral buoyancy multiple times during a single dive make it less suited for certain types of midwater biological research than certain non-NDSF assets. For this reason, arguments favoring the full utilization of NDSF assets have the unintended consequence of restricting the scope of deep submergence science.

A reasonable solution to this problem is to upgrade the capabilities of NDSF assets so that they can be used in all fields of deep submergence science. This would expand the scope of deep submergence science and maintain costs at a reasonable level. Although this is clearly an important part of the solution, by itself, it may be inadequate. First, these upgrades, if they occur, will not be completed for two to three years, and some short-term measures are needed. Perhaps more importantly, there is a danger that the existing pattern of use of NDSF assets will simply persist. One way to address both of these problems is for NSF/OCE to provide modest, but immediate, funding to support the use of non-NDSF assets (NOAA currently funds the use of non-NDSF facilities on a modest basis). This funding should not be drawn from the NSF/OCE science program budgets but should be allocated by the NSF/OCE Integrative Facilities Program. An additional benefit of establishing such a fund is that it would provide a gauge of demand for capabilities not provided by current NDSF assets.

Recommendation: NSF/OCE should establish a small pool of additional funds (on the order of 10 percent of the annual budget for NDSF) that could be targeted specifically to support the use of non-NDSF vehicles for high-quality, funded research, when legitimate barriers to the use of NDSF assets (as opposed to personal preference) can be demonstrated.

If, as additional assets become available (through either purchase or construction), the demand for non-NDSF vehicles declines (or never materializes), these funds could be used to address other (non-deep submergence) marine operational needs as determined by NSF/OCE.

DEVELOPING NEW ASSETS

Reforming the asset management system to allow for wider (though still limited) access to deep submergence assets will, by itself, not be sufficient to meet the needs of science. Existing assets are simply too limited in their capabilities and capacity, especially at depths greater than 3,000m, to support the growing demand to conduct research over the necessary geographic and depth range. High demand for existing deep-diving assets within the NDSF pool has forced asset managers to place a heavy premium on maximizing operational days while minimizing days in transit. The pressure that this geographic restriction has led to can only be expected to increase as ongoing efforts to address a more scientifically diverse set of problems increase the demand for deep-diving vehicles to work in diverse settings.

Recommendation: NSF/OCE should construct an additional scientific ROV system dedicated to expeditionary research,[2] to broaden the use of deep submergence tools in terms of the number of users, the diversity of research areas, and the geographical range of research activities.

Furthermore, while such an ROV system can be constructed using well-established subsystems several factors should be considered during its design. Probably most important overall is the incorporation of several features that will greatly enhance its utility as well as its ability to complement existing assets. Some of the most significant features include a robust variable ballast system, standardized tooling suites, open software and hardware architectures, electronic thruster systems, tether management systems, improved handling systems, and camera and lighting system. The total cost of this system would be approximately $5 million, and it could be built and ready for service within one year of authorization. Using the current University-National Oceanographic Laboratory System (UNOLS) model for marine operations, this ROV system could be mobilized onto the current fleet without any significant addition of hardware. The operational requirements would be similar to those for the current *Jason II* crew. The operational costs of this new ROV should be similar to those of *Jason II* and, thus, would represent a 20 percent increase in the overall operating costs of NDSF. This increase should have a modest impact if it is anticipated and the overall budget is increased incrementally in preparation for the construction and operation of a new ROV. One justification for adding a new ROV system to the NDSF asset pool is to provide even greater geographic range to the growing number of ocean scientists seeking access to deep submergence assets.

[2]As opposed to those needed to support the Ocean Observatories Initiative.

Recommendation: NSF/OCE should, after a proper analysis of the cost-benefits of distributed facilities, strongly consider basing this new ROV system at a second location that would minimize the transit time for periodic overhaul and refit of both ROV systems.

The best approach to deep submergence science is the use of a combination of tools. Detailed reconnaissance surveys are best achieved using tethered vehicles and AUVs. Experiments and observatory work that require longer time at already well characterized sites on the seafloor are best conducted with ROVs. Moreover, work at depths greater than 6,500m will definitely require unoccupied vehicles, as long as the expense and risk of constructing and operating HOVs capable of work at these depths discourage their use.

As discussed in Chapter 3, human presence at depth remains a significant lynchpin in the nation's oceanographic research effort. Detailed descriptions of specific sites or work in the water column benefit from the direct human observation allowed by HOVs. Despite rapid and impressive growth in the capabilities of unoccupied vehicles (both remotely operated and autonomous), the scientific demand for HOV access can be expected to remain high. However, the capabilities of the existing *Alvin* limit its scientific usefulness for some types of deep ocean research. Improving these capabilities, even without extending its depth range, is clearly necessary if many of the high-priority scientific goals discussed in Chapter 2 are to be achieved.

Recommendation: NSF/OCE should construct a new, more capable HOV (with improved visibility, neutral buoyancy capability, increased payload, extended time at working depth, and other design features discussed in Chapter 4).

The bulk of existing *Alvin* use is at depths considerably shallower than its 4,500-m limit. Even at these shallower depths, scientific demand remains unmet. At the same time, certain scientific goals would be furthered by the acquisition of an HOV with a 6,500-m-depth range. Under current safety guidelines imposed by the operating institutions, *Alvin* and other HOVs are prohibited from operating in waters deeper than the rated working depth, even if this operation is in the water column. For this reason alone, providing access to an HOV with a greater depth capability would allow its use over a broader geographical range, thereby improving its utility for a portion of the potential user community. As discussed at length in Chapter 4 however, it is not clear at present that a suitable sphere can be obtained to allow the fabrication of a deeper-diving HOV, especially given the limited funds available to NSF/OCE in the next two fiscal years.

The most promising approaches for moving ahead during the time frame articulated by NSF/OCE would make use of one of two existing spheres. The first is the third, unused sphere from the Russian *Mir* HOV series (referred to as the Lokomo sphere), which has been rated to 6,500m. The other available sphere is the titanium sphere used in the existing *Alvin*, which is rated to 4,500m. While the possibility of fabricating an entirely new sphere warrants investigation, there is insufficient information at this time to determine the ultimate availability and cost of a specifically fabricated sphere. Given the technical and cost uncertainties, and that the scientific justification for conducting HOV operations at depths greater than 4,500m appears to be incremental (i.e., it represents promising but logical extensions of work supported at shallower depths), it is not clear that significant additional resources (i.e., in excess of those needed to fully upgrade the current NDSF HOV capability;, as discussed in Chapter 4) should be expended on a new HOV with greatly extended depth capability if that expenditure were to preclude construction of the ROV system recommended in Chapter 4 of this report.

Recommendation: Thus, constructing an HOV capable of operating at significantly greater depths (6,000m plus) should be undertaken only if additional design studies demonstrate that this capability can be delivered for a relatively small increase in cost and risk as discussed in Chapter 4.

To implement these recommendations, NSF and other NDSF sponsors would have to increase funding at a rate of 10 to 15 percent over the next three years to (1) cover the cover the cost of non-NDSF vehicle use and (2) cover the cost of the new ROV. In order to provide the capabilities and capacity to meet existing and anticipated demands, NSF and other NDSF sponsors should take a three-step approach: (1) set aside additional funds called for to support non-NDSF vehicle use as quickly as possible; (2) initiate acquisition of the new ROV in 2004 or 2005; and (3) undertake a detailed engineering study to evaluate various HOV enhancement options called for in Chapter 4 with an aim of delivering these new platforms by 2006. It is entirely possible, perhaps even probable, that given more time and significantly greater funds, the federal agencies that fund deep submergence research could build a number of platforms with greater capabilities than those described here. The statement of task, however, was specifically crafted to ensure that advice provided in this report was appropriate given the current fiscal and programmatic realities facing federal science agencies. If, in the future, these requirements were to change significantly, then the appropriate mix of assets needed to support the deep submergence effort should be revisited.[3]

[3]The purpose of this study is to provide NSF with recommendations for consideration regarding activities to provide infrastructure support for basic research at depth in the oceans through NDSF or other means. As such, the discussions in this report are designed to inform this question and are not intended to provide an exhaustive account of all research-related activities carried out at depth or a complete account of all the potential assets that exist. The discussion of assets in this report is limited therefore to those that establish whether adequate deep submergence vehicles exist within or outside the National Deep Submergence Facility. Furthermore, any recommendations made in this report are above and beyond the needs for other large programs such as NSF's Ocean Observatories Initiative or activities falling within the realm of ocean exploration.

References

Behl, R.J., and J.P. Kennett. 1996. Brief interstadials events in Santa Barbara Basin, Northeast Pacific, during the past 60 kyr. *Nature* 379:243-246.

Booth, J.S., W.J. Winters, and W.P. Dillon. 1994. Circumstantial evidence of gas hydrate and slope failure associations on the U.S. Atlantic continental margin. Pp. 487-489 in E.D. Sloan, Jr., J. Happel, and M.A. Hnatow, eds. *International Conference on Natural Gas Hydrates: Annals of the New York Academy of Sciences* held in New Paltz, New York, June 20, 1993. New York Academy of Sciences, New York.

Brierly, A., P. Fernandez, M. Brandon, F. Armstrong, N. Millard, S. McPhail, P. Stevenson, M. Pebody, J. Perrett, M. Squires, D. Bone, and G. Griffiths. 2002. Antarctic krill under sea ice: Elevated abundance in a narrow band just south of ice edge. *Science* 295:1890-1892.

Brown, R.S., D. Foster, and B. Walden. 2000. *Comparison of Options for an Improved Manned Deep Submersible Vehicle for Research*. Woods Hole Oceanographic Institution Proposal No. 2663. Woods Hole Oceanographic Institution. Woods Hole, MA.

Cable, J.E., W.C Burnett, J.P. Chanton, and G.L. Weatherly. 1996. Estimating groundwater discharge into the northeastern Gulf of Mexico using radon-222. *Earth and Planetary Science Letters* 144:591-604.

Carson, B., E. Suess, and J.C Strasser. 1990. Fluid flow and mass flux determinations at vent sites on the Cascadia margin accretionary prism. *Journal of Geophysical Research* 95:8891-8897.

Corliss, J.B., J. Dymond, L.I. Gordo, J.M. Edmond, R.P. von Herzen, R.D. Ballard, K. Green, D. Williams, A. Bainbridge, K. Crane, and T.H. van Andel. 1979. Submarine thermal springs on the Galapagos Rift. *Science* 203:1073-83.

Dickens, G.R., and M.S. Quinby-Hunt. 1997. Methane hydrate stability in pore water: A simple theoretical approach for geophysical applications. *Journal of Geophysical Research* 102:773-783.

Druffel, E.R., and B.H. Robison. 1999. Is the deep sea on a diet? *Science* 284:1139-1140.

Federal Oceanographic Facilities Committee, National Oceanographic Partnership Program. 2001. *Charting the Future for the National Academic Research Fleet: A Long Range Plan for Renewal*. A Report from the Federal Oceanographic Facilities Committee (FOFC) of the National Oceanographic Partnership Program (NOPP) to the National Ocean Research Leadership Council (NORLC). National Oceanographic Partnership Program Office, Washington, DC.

Firoozfam, P., and S. Negahdaripour. 2002. *Application of panoramic images of the seafloor for improved motion estimation and photo-mosaicing*. Presented to the MTS/IEEE Oceans Conference held in Biloxi, MS, October 28-31, 2002. Institute of Electrical and Electronic Engineers, Piscataway, NJ.

Firoozfam, P., and S. Negahdaripour. 2003. A multi-camera conical imaging system for robust 3-D motion estimation, positioning and mapping from UAVs. Pp. 99-106 in *Proceedings of the IEEE Conference on Advanced Video and Signal Based Surveillance* held in Miami, FL, July 21-22, 2003. Institute of Electrical and Electronic Engineers, Piscataway, NJ.

Fryer, P. 1992. Mud volcanoes of the Marianas. *Scientific American* 266:46-52.

Gluckman J., and S.K. Nayar. 2000. Rectified catadioptric stereo sensors. *IEEE Transactions on Pattern Analysis and Machine Intelligence* 24(2):224-236.

Handa, Y.P. 1990. Effect of hydrostatic pressure and salinity on the stability of gas hydrates. *Journal of Physical Chemistry* 94:2652-2657.

Herring, P. 2002. *The Biology of the Deep Ocean*. Oxford University Press, University of Oxford, UK.

Kastner, M., H. Elderfield, and J.B. Martin. 1991. Fluids in covergent margins: What do we know about their composition, origin, role in diagenesis and importance for oceanic chemical fluxes? *Philosophical Transactions of the Royal Society of London*, Series A 335:243-259.

Kastner, M., H. Elderfield, W.J. Jenkins, J.M. Gieskes, and T. Gamo. 1993. Geochemical and isotopic evidence for fluid flow in the western Nankai subduction zone, Japan. *Proceedings of the Ocean Drilling Program, Scientific Results* 131:397-413.

Kayen, R.E., and H.J. Lee. 1991. Pleistocene slope instability of gas hydrate-laden sediment on the Beaufort Sea margin. *Marine Geotechnology* 10:125-141.

Knipe, R.J., and A.M. McCaig. 1994. *Microstructural and Microchemical Consequences of Fluid Flow in Deforming Rocks*. Special Publication, 78:99-111. Geological Society of London, England.

Kulm, L.D., E. Suess, J.C. Moore, B. Carson, B.T. Lewis, S.D. Ritger, D.C. Kadko, T.M. Thornburg, R.W. Embley, W.D. Rugh, G.J. Massoth, M.G. Langseth, G.R. Cochrane, and R.L. Scamman. 1986. Oregon subduction zone: Venting, fauna, and carbonates. *Science* 231:561-566

Kvenvolden, K.A. 1988. Methane hydrate—A major reservoir of carbon in the shallow geosphere? *Chemical Geology* 71:41-51.

Kvenvolden, K.A. 1995. A review of the geochemistry of methane in natural gas hydrate. *Organic Geochemistry* 23:997-1008.

Langseth, M., and J.C. Moore. 1990. Fluids in accretionary prisms. *Eos* 71:235, 245-246.

MARGINS. 1998. SEIZES Science Plan. MARGINS. [Online]. Available at: http://www.margins.wustl.edu/SciPlan.html. [2003, October 8].

MARGINS. 2000a. Subduction Factory Science Plan. MARGINS. [Online]. Available at: http://www.margins.wustl.edu/SciPlan.html. [2003, October 13].

MARGINS. 2000b. Source to Sink Science Plan. MARGINS. [Online]. Available at: http://www.margins.wustl.edu/SciPlan.html. [2003, October 13].

Marshall, N.B. 1979. *Deep-Sea Biology, Developments and Perspectives*. Garland Press, New York.

McCutchen, D., inventor. 1997. Dodeca, LLC, assignee. December 30, 1997. Immersive dodecaherdral video viewing system. U.S. Patent 5,703,604.

Moore, J.C., and P. Vrolijk. 1992. Fluids in accretionary prisms. *Reviews of Geophysics* 30:113-135.

Nalwa, V., inventor. 1996. AT&T, assignee. July 23, 1996. Panoramic projection apparatus. U.S. 5,539,483.

National Research Council (NRC). 1996. *Undersea Vehicles and National Needs*. National Academy Press, Washington, DC.

National Research Council (NRC). 2000. *Illuminating the Hidden Planet*. National Academy Press, Washington, DC.

National Research Council (NRC). 2003a. *Enabling Ocean Research in the 21st Century: Implementation of a Network of Ocean Observatories*. National Academies Press, Washington, DC.

National Research Council (NRC). 2003b. *Exploration of the Seas: Voyage into the Unknown*. National Academies Press, Washington, DC.

Nayar, S.K. 1997. Catadioptric omindirectional camera. Pp. 482-488 in *Proceedings of the IEEE Conference of Computer Vision Pattern Recognition* held in San Juan, PR, June 17-19, 1997. Institute of Electrical and Electronic Engineers, Piscataway, NJ.

Negahdaripour, S., H. Zhang, P. Firoozfam, and J. Oles. 2001. *Utilizing Panoramic Views for Visually Guided Tasks in Underwater Robotics Applications*. Presented to the MTS/IEEE Oceans Conference held in Honolulu, HI, November 6-8, 2001. Institute of Electrical and Electronic Engineers, Piscataway, NJ.

Neumann, J., C. Fermuller, and Y. Aloimonos. 2001. *Eyes from Eyes: New Cameras for Structure from Motion*. Center for Automation Research Technical Report. University of Maryland, College Park.

Oliver, J. 1986. Fluids expelled tectonically from orogenic belts: Their role in hydrocarbon migration and other geologic phenomena. *Geology* 14:99-102.

Peacock, S.M. 1990. Fluid processes in subduction zones. *Science* 248:329-337.

Peleg, S., M. Ben-Ezra, and Y. Pritch. 2001. OmniStereo: Panoramic stereo imaging. *IEEE Transactions on Pattern Analysis and Machine Intelligence* 23(3):279-290.

Raynaud D., J. Jouzel, J.M. Barnola, J. Chappellaz, R.J. Delmas, and C. Lorius. 1993. The ice record of greenhouse gases. *Science* 259:926-934

RIDGE 2000. 2003. *RIDGE 2000 Science Plan*. RIDGE 2000. [Online]. Available at: http://ridge2000.bio.psu.edu/R2k_sci_plan_2003.pdf. [2003, October 8].

Robison, B.H. 2000. The coevolution of undersea vehicles and deep-sea research. *Marine Technology Society Journal* 33:65-73.

Rona, P.A. 2001. Deep-diving manned research submersibles. *Marine Technology Society Journal* 33(4):13-25.

Rothschild, L.J., and R.L. Mancinelli. 2001. Life in extreme environments. *Nature* 409:1092-1101.

Sheridan, T.B. 2002. *Humans and Automation: System Design and Research Issues*. Wiley Interscience, Hoboken, NJ.

Sloan, E.D. 1990. *Clathrate Hydrates of Natural Gases*. Marcel Dekker, New York, NY.

Smith, K.L., Jr., and R.S. Kaufmann. 1999. Long-term discrepancy between food supply and demand in the deep eastern north Pacific. *Science* 284(5517):1174-1177.

Smith, K.L., R.C. Glatts, R.J. Baldwin, S.E. Beaulieu, A.H. Uhlman, R.C. Horn, and C.E. Reimers. 1997. An autonomous, bottom-transecting vehicle for making long time-series measurements of sediment community oxygen consumption to abyssal depths. *Limnology and Oceanography* 42:1601-1612.

Swaminathan, R., and S.K. Nayar. 2000. Nonmetric calibration of wide-angle lenses and polycameras. *IEEE Transactions on Pattern Analysis and Machine Intelligence* 22(10):1172-1178.

University-National Oceanographic Laboratory System (UNOLS). 1990. *Submersible Science Study for the 1990s*. Report funded by the National Science Foundation, Office of Naval Research, and National Oceanographic and Atmospheric Administration. UNOLS Office, University of Rhode Island, Narragansett, RI.

University-National Oceanographic Laboratory System (UNOLS). 1994. *The Global Abyss: An Assessment of Deep Submergence Science in the United States.* UNOLS Office, University of Rhode Island, Narragansett, RI.

University-National Oceanographic Laboratory System (UNOLS). 1999. *Descend '99 Proceedings* held in Arlington, VA, October 25-27, 1999. UNOLS Office, University of Rhode Island, Narragansett, RI.

U.S. Air Force Scientific Advisory Board. 2003. *Unmanned Aerial Vehicles in Perspective: Effects, Capabilities, and Technologies*. SAB-TR-03-01. [Online] Available at: https://www.sab.hq.af.mil/2003_studies/studies.html. [2003, September 8]

Widder, E.A. 1997. Bioluminescence. *Sea Technology* (March):33-39.

Widder, E.A. 2003. *Future Deep Submergence Needs for Midwater Biology*. Presentation to the National Research Council, Ocean Studies Board, Committee on Future Needs in Deep Submergence Science, held in San Francisco, CA, June 25-26, 2003. National Research Council, Washington, DC.

Yoerger, D., S. Bradlet, and B. Walden. 1998. The Autonomous Benthic Explorer (ABE): A deep ocean AUV for scientific seafloor survey. *Sea Technology* 33:50-54.

Appendix A

Committee and Staff Biographies

John A. Armstrong *(Chair)* received his Ph.D. in the field of nuclear magnetic resonance from Harvard University in 1961. Dr. Armstrong spent most of his career at IBM, until he retired as vice president of science and technology. He is the author or coauthor of some 60 papers on nuclear resonance, nonlinear optics, the photon statistics of lasers, picosecond pulse measurements, the multiphoton spectroscopy of atoms, the management of research in industry, and issues of science and technology policy. As a result of his contributions in nonlinear optics, quantum physics, and technical leadership in advanced very-large-scale integration technology, Dr. Armstrong was elected a member of the National Academy of Engineering (NAE) in 1987. In addition, he received the George E. Pake Prize of the American Physical Society in 1989. Dr. Armstrong was a member of the presidentially appointed National Advisory Committee on Semiconductors. He was also a member of the National Science Board from 1996 to 2002 and served on its Special Commission on the Future of the National Science Foundation (NSF). Dr. Armstrong has served on numerous National Research Council (NRC) bodies, including the Commission on Physical Sciences, Mathematics, and Applications, where he was liaison to the Computer Science and Technology Board; he chaired the Committee on Partnerships in Weather and Climate Services.

Keir Becker obtained his Ph.D. in oceanography from Scripps Institution of Oceanography in 1981. He is a professor of marine geology and geophysics at the University of Miami. Dr. Becker's research interests include heat flow and hydrothermal circulation in the oceanic crust, permeability and other physical properties of oceanic crust, and borehole hydrogeological observatories. He has published numerous papers on oceanic heat flow and downhole experiments in oceanic crust. Over the past 10 years, Dr. Becker has sailed on various scientific expeditions involving the Ocean Drilling Program (ODP) drillship, the deep submergence vehicle (DSV) *Alvin*, the French submersible *Nautile*, and the Japan Marine Science and Technology Center (JAMSTEC) remotely operated vehicle (ROV) *Kaiko*. He chairs the Joint Oceanographic Institutions for Deep Earth Sampling Science Committee and recently co-chaired the Dynamics of Earth and Ocean Systems Steering Committee. He has served on many other panels, such as the NRC Committee on Sea Floor Observatories and the Woods Hole Oceanographic Institution New Alvin Design Advisory Committee.

Thomas W. Eagar received his Sc.D. from the Massachusetts Institute of Technology (MIT) in 1975 in metallurgy. Currently, he is a Thomas Lord Professor of Materials at MIT, where his research focuses on many aspects of metal fabrication, including the fundamentals of bonding composites, superalloys, and electronic packaging as well as improved methods of dimensional analysis of materials processing. With considerable experience in design analysis and fabrication of submarines and other naval vessels, Dr. Eagar was named a member of NAE in 1997 and a fellow and honorary member of the American Welding Society. He has served on a number of NRC committees including the National Materials Advisory Board, the Committee for Investigation of Steels for Improved Weldability in Ship Construction, and the Committee on New Directions in Manufacturing.

Bruce Gilman earned his bachelor's in aeronautical engineering from the Polytechnic University of New York in 1960. He recently retired as president and chief executive officer of Sonsub, Inc. His experience spans four decades of the offshore industry with particular emphasis on subsea intervention engineering and operations. He has been an active participant in the development of the industry from the earliest days of manned surface and bell-supported diving through human-occupied submersibles up to today's most advanced ROV systems. He has participated in offshore operations, designed and developed equipment, and served as senior executive with some the most preeminent subsea engineering and services contractors and equipment manufacturers. Mr. Gilman is a registered professional engineer and holds several patents relating to the offshore industry. He is also a Marine Technology Society fellow and a member of the American Society of Mechanical Engineers and Society of Petroleum Engineers. Mr. Gilman serves on the Texas Sea Grant College Program Advisory Committee and on the NRC Committee on Exploration of the Seas and recently completed an assignment as a peer-review panelist on the 2003 National Oceanic and Atmospheric Administration (NOAA) ocean exploration proposal review panel.

Mark Johnson earned his M.S. in ocean engineering from the Florida Institute of Technology in 1991. He is the lead subsea engineer as a consultant for BP Deepwater Production GoM and performs project management of deepwater subsea interventions. Mr. Johnson's BP project management experience encompasses reviewing current strategies for autonomous underwater vehicles (AUVs) and developing and approving new tooling for deepwater projects including diver, ROV, and 1-atmosphere interventions. His research interests include design and construction of a surf-zone ROV for the U.S. Army Corps of Engineers, hyperbaric medicine, and design of life-support systems. Mr. Johnson's company, O2 Dive Technologies, designs and manufactures rebreather systems for the National Aeronautics and Space Administration (NASA) International Space Station, emergency response applications, and technical diver use. He is also a founding member of Rebreather Technologies Inc., a new diving agency for the certification of technical and rebreather divers.

Miriam Kastner earned her Ph.D. in geology in 1970 from Harvard University. She is a professor of earth sciences in the Graduate Department of Scripps Institution of Oceanography at the University of California, San Diego. Dr. Kastner's expertise is in marine geochemistry and her research focuses on chemical paleoceanography; the role and fluxes of fluids in convergent plate margins; the origin, environmental implications, and diagenesis of primarily marine authigenic minerals (i.e. phosphates, sulfates, silicates, carbonates); gas hydrates in continental margins and implications for global change; and submarine hydrothermal deposits. Her publications cover areas in gas hydrates, fluid flow paths, oceanic minerals, and hydrothermal deposits. She is a fellow of the American Geophysical Union, Geochemical Society, and the American Assocation for the Advancement of Science (AAAS). She has sailed on a number of scientific expeditions involving submersibles and ROVs, as well as on the ODP. Dr. Kastner serves on the NRC Ocean Studies Board.

Dhugal John Lindsay received his Ph.D. in aquatic biology from the University of Tokyo in 1998. He is a research scientist with the Japan Marine Science and Technology Center. Dr. Lindsay's research focuses on midwater ecology, particularly concentrating on gelatinous organisms that are too fragile to be sampled by conventional methods. Dr. Lindsay has extensive experience with the Japanese research vessel and submersible fleet, both as chief scientist and as a member of multidisciplinary teams. His sailing experience includes more than 35 cruises aboard various Japanese research vessels and 19 dives in crewed submersibles. He has used conventional sampling techniques such as nets and sediment traps (e.g., *R/V Tanseimaru,* University of Tokyo) and towed camera arrays (e.g., 4,000m and 6,000m Deep-Tow Cameras, R/V *Kaiyo*) and has also used both manned submersibles (e.g., *Shinkai 2000, R/V Natsushima; Shinkai 6500, R/V Yokosuka*) and remotely-operated vehicles (e.g., ROV *Dolphin* 3K, *R/V Natsushima;* ROV *Ventana, R/V Point Lobos;* ROV *HyperDolphin, R/V Kaiyo;* ROV *Kaiko, R/V Kairei*) to investigate fauna from depths as shallow as the euphotic layer to as deep as the Challenger Deep, Mariana Trench. Dr. Lindsay is a member of the Japanese Society of Fisheries Science, Plankton Society of Japan, and Oceanographic Society of Japan; is on the editorial board of the journal *Plankton Biology and Ecology;* and is currently serving on the interim planning committee for the Okinawa Marine Life Science Research Institute.

Catherine Mevel received her Ph.D. in 1975 from the Université Pierre et Marie Curie, Paris. She is the deputy director of the Laboratoire de Géosciences Marines (LGM) at Institut de Physique du Globe de Paris. Dr. Mevel's scientific interests concern the interaction between seawater and the oceanic lithosphere; they include paths of seawater penetration, physical conditions of interaction, and mineralogical and chemical consequences. Her experience at sea encompasses participation on the PHARE cruise with the ROV *Victor,* a number of cruises with the submersible *Nautile* and *Cyana,* and the INDOYO cruise with the submersible *Shinkai 6500*. She has also been involved in many mapping and sampling cruises. Dr. Mevel chaired the Dorsales program, the ODP-France, and was the French representative at ODP EXCOM. She is a member of the InterRidge Steering Committee and the Scientific Committee of the French Research Institute for the Exploration of the Sea.

Shahriar Negahdaripour earned his Ph.D. from MIT in 1987. He is a professor in the Electrical and Computer Engineering Department of the University of Miami. Dr. Negahdaripour's projects involve underwater vision and imaging; they include three-dimensional shape recovery from image shading for automatic computer recognition of underwater objects, real-time PC-based vision system for seafloor image mosaicing, automatic optical station keeping, navigation of underwater robotic vehicles, adaptive optical sensing for vision-based three-dimensional target recognition from underwater images, and motion-based video compression for underwater application. He is a senior member of the Institute of Electrical and Electronics Engineers (IEEE) and has served as the co-chair of the IEEE Computer Society International Symposium on Computer Vision and the IEEE Computer Society Conference on Computer Vision Pattern Recognition.

Shirley A. Pomponi holds a Ph.D. in biological oceanography from the University of Miami conferred in 1977. She is currently the vice president and director of research at Harbor Branch Oceanographic Institution. Her research focuses on the development of methods for sustainable use of marine resources for drug discovery and development and, in particular, on the cell and molecular biology of sponges with biomedical importance. Dr. Pomponi has led numerous research expeditions worldwide. She is a member of the Society for In Vitro Biology, the American Geophysical Union, and the American Society for Cell Biology. Dr. Pomponi served on the President's Panel on Ocean Exploration and on the NRC's Committees on Marine Biotechnology: Development of Marine Natural Products and Oceans and Human Health Panel. She is currently serving on the NRC's Committee on Exploration of the Seas, the Scientific Advisory Panel to the U.S. Commission on Ocean Policy, and the Ocean Studies Board.

Bruce Robison received his Ph.D. in biological oceanography from Stanford University in 1973. He is the senior scientist and former science chair of the Monterey Bay Aquarium Research Institute. Dr. Robison's research interests are focused on deep-sea ecology and applying advanced submersible technology to oceanographic research. A qualified manned submersible pilot, he has dived in 12 different research submersibles and is a regular user of ROVs. Dr. Robison led the Deep Rover expedition, the first program to use submersibles to study California's Monterey Submarine Canyon, in 1985. He is a fellow of AAAS and the California Academy of Sciences. He received the Marine Technology Society's Lockheed Martin Award for Ocean Science and Engineering and was a recipient of the Monterey Bay National Marine Sanctuary, Science/Research Award. Dr. Robison has served on the NRC Committee on Undersea Vehicles and National Needs.

Andrew Solow earned his Ph.D. in geostatistics from Stanford University in 1986. Dr. Solow is an associate scientist and director of the Marine Policy Center at Woods Hole Oceanographic Institution. His research experience involves environmental statistics, time-series analysis, spatial statistics, Bayesian methods, statistical biology, and ecology. Dr. Solow has authored or coauthored some 120 scientific publications on topics that range from biological diversity, El Niño, to empirical analysis on volcanic eruptions. In addition to his work in environmental and ecological statistics, he has worked on problems connected to the value of scientific information. Dr. Solow is a former member of the NRC's Commission on Geosciences, Environment, and Resources and the Committee on Fifty Years of Ocean Discovery at the National Science Foundation. He is a current member of the Committee for Review of the U.S. Climate Change Science Program Strategic Plan.

Greg Zacharias obtained his Ph.D. in aeronautics and astronautics from MIT in 1977. He is a senior principal scientist at Charles River Analytics; and leads efforts in human behavior modeling and agent-based decision support systems. Before founding Charles River, Dr. Zacharias was a senior scientist at BBN Technologies, a research engineer at C.S. Draper Labs, and a United States Air Force attaché for the Space Shuttle program at NASA Johnson Space Center. He has been a member of the NRC Committee on Human Factors since 1995 and served on the NRC Panel on Modeling Human Behavior and Command Decision Making. Dr. Zacharias is a member of the Department of Defense Human Systems (HS) Technology Area Review and Assessment Panel, a member of the USAF Scientific Advisory Board, and chairman of the USAF Human System Wing Advisory Group for Brooks Air Force Base.

Staff

Dan Walker *(Study Director)* obtained his Ph.D. in geology from the University of Tennessee in 1990. A senior program officer at the Ocean Studies Board, Dr. Walker also holds a joint appointment as a guest investigator at the Marine Policy Center of the Woods Hole Oceanographic Institution. Since joining the Ocean Studies Board in 1995, he has directed a number of studies including *Environmental Information for Naval Warfare* (2003); *Oil in the Sea III: Inputs, Fates, and Effects* (2002); *Clean Coastal Waters: Understanding and Reducing the Effects of Nutrient Pollution* (2000); *Science for Decisionmaking: Coastal and Marine Geology at the U.S. Geological Survey* (1999); *Global Ocean Sciences: Toward an Integrated Approach* (1998); and *The Global Ocean Observing System: Users, Benefits, and Priorities* (1997). A member of the American Geophysical Union, the Geological Society of America, and the Oceanography Society, Dr. Walker was recently named editor of the *Marine Technology Society Journal*. A former member of both the Kentucky and the North Carolina Geologic Surveys, Dr. Walker's interests focus on the value of environmental information for policy making at local, state, and national levels.

Joanne Bintz earned her Ph.D. in biological oceanography from the University of Rhode Island Graduate School of Oceanography. Dr. Bintz has conducted research on the effects of decreasing water quality on eelgrass seedlings and the effects of eutrophication on shallow macrophyte-dominated coastal ponds using mesocosms. She has directed NRC studies on *The Review of the Florida Keys Carrying Capacity*; *Chemical Reference Materi-*

als: Setting the Standard for Ocean Science; and *Enabling Ocean Research in the 21st Century: Implementation of a Network of Ocean Observatories.* Her interests include coastal ecosystem ecology and restoration, marine technology, oceanographic education, and coastal management and policy.

John Dandelski received his M.A. in marine affairs and policy from the Rosenstiel School of Marine and Atmospheric Science, University of Miami, where his research focused on evaluating fisheries' impacts on the benthic communities of Biscayne Bay and where he served as the school's assistant diving safety officer. He has been with the National Academies since 1998 and with the Ocean Studies Board as a research associate since 2001. As a graduate research intern at the Congressional Research Service he wrote reports for Congress on fisheries and ocean health issues. Mr. Dandelski is currently study director for the River Basins and Coastal Systems Panel of the Committee to Assess the U.S. Army Corps of Engineers Methods of Analysis and Peer Review for Water Resources Project Planning, was the project manager for *Abrupt Climate Change: Inevitable Surprises,* and has worked on a number of other reports including *Environmental Information for Naval Warfare* and *Oil in the Sea III: Inputs, Fates, and Effects.* Mr. Dandelski also holds an M.S. in industrial and organizational psychology, and his interests include environmental experiential education, information systems, diving health and safety, and marine policy.

Sarah Capote earned her B.A. in history from the University of Wisconsin-Madison in 2001. She is a project assistant with the Ocean Studies Board. During her tenure with the Board, Ms. Capote assisted with the completion of the studies on *Nonnative Oysters in the Chesapeake Bay* and *Exploration of the Seas: Voyage into the Unknown.*

Appendix B

Acronyms

ABE	Autonomous Benthic Explorer
ABS	American Bureau of Shipping
AUV	autonomous underwater vehicle
BBC	British Broadcasting Corporation
CCD	charge-coupled device
CoML	Consensus of Marine Life
COTS	commercial off-the-shelf
CPU	central processing unit
CSSF	Canadian Scientific Submersible Facility
DESCEND	DEveloping Submergence SCiencE for the Next Decade
DESSC	Deep Submergence Science Committee
DP	dynamic positioning
DSV	Deep Submergence Vehicle
DV	digital video
FOV	field of view
GTAW	gas tungsten arc welding
HARP	High-Gain Avalanche Rushing Photoconductor
HBOI	Harbor Branch Oceanographic Institution
HCI	human-computer interface
HD	high definition
HDTV	high-definition television
HOV	Human Occupied Vehicles

H-SI	Human-System Integration
HURL	Hawaii Undersea Research Laboratory
IFREMER	French Research Institute for Exploration of the Seas
JAMSTEC	Japan Marine Science and Technology Center
JSL	*Johnson Sea-Links I* and *II*
LCD	liquid crystal display
MBARI	Monterey Bay Aquarium Research Institute
MG&G	marine geology and geophysics
MOA	Memorandum of Agreement
NDSF	National Deep Submergence Facility
NMR	Nuclear Magnetic Resonance
NOAA	National Oceanic and Atmospheric Administration
NRC	National Research Council
NSF	National Science Foundation
NURP	NOAA's Undersea Research Program
OCE	NSF's Ocean Sciences Division
ONR	Office of Naval Research
OOI	NSF's Ocean Observatories Initiative
OS	operating system
PI	principal investigator
RIDGE	Ridge Interdisciplinary Global Experiments
ROV	remotely operated vehicle
SC	spatially correspondent
SONAR	sound navigation and ranging
SSSC	Submersible Science Study Committee
SWATH	small waterplane area twin hull
TMS	tether management system
UAV	unmanned aerial vehicle
UNOLS	University-National Oceanographic Laboratory System
VR	virtual reality
WHOI	Woods Hole Oceanographic Institution

Appendix C

International Autonomous Underwater Vehicles Listing

Institution	Major Topics	Vehicles
University of Aberdeen Ocean Research Lab Scotland	Autonomous landers and acoustics	Aberdeen University Deep Ocean Submersible
Alfred Wegener Institute Deepsea Research Bremerhaven, Germany	Autonomous landers	Autonomous underwater vehicle (AUV) payload modules
Autonomous Undersea Systems Institute Marine Systems Engineering Laboratory Lee, New Hampshire	Environmental monitoring, generic behaviors, and control	AUVs
Australian National University Robotics Systems Laboratory Canberra	Underwater exploration and observation	*Kambara*
Bluefin Robotics Corp. Cambridge, Massachusetts	AUVs	*Odyssey I*, *Odyssey II B*, *Odyssey III*, and *Seasquirt*
C & C Technologies, Inc. Lafayette, Louisiana	AUVs and survey services	*Hugin 3000*

Institution	Major Topics	Vehicles
Instituto Automazione Navale Consiglio Nazionale delle Ricerche Robot Lab Genova, Italy	Control, navigation, and manipulation	*Romeo* and *Aramis*
Technical University of Denmark Department of Automation Lyngby, Denmark	Sonar for underwater inspection	*Martin*
Instituto Superior Tecnico Dynamical Systems and Ocean Robotics Laboratory Lisbon, Portugal	Installations, long-range missions, exploration, and control	*Caravela*, Marine Utility Vehicle System, and *Sirene*
University of Florida Machine Intelligence Laboratory Gainesville	AUVs for competitions	*SubjuGator*
Florida Atlantic University Advanced Marine Systems Laboratory Boca Raton		*Ocean Voyager II, Ocean Explorer*, and *Bottom Classification* and *Albedo Package (BCAP)*
Hafmynd Ltd. Reykjavik, Iceland	AUVs	*Gavia*
Harbor Branch Oceanographic Institution Ocean Engineering and Production Division Fort Pierce, Florida	AUVs	*Ocean Voyager*
University of Hawaii Autonomous Systems Laboratory Honolulu	Navigation, search, and recognition	*Omni-Directional Intelligent Navigator*
Heriot-Watt University Ocean Systems Laboratory Edinburgh, Scotland	Vision, sonar, manipulation, simulation, acoustics, electromagnetic and optical communication, positioning, navigation, and sampling	*Autonomous Light Intervention Vehicle, Aramis*, and *Rauver*

Institution	Major Topics	Vehicles
Hyland Underwater Vehicles Edinburgh, Scotland	Simple, small, proof-of-concept AUV	*MicroSeeker*
French Institute of Research and Exploration of the Seas Data Processing Systems Toulon	Control and navigation and control architectures	Open and Reconfigurable Vehicle for Experimental Techniques
International Submarine Engineering Ltd. Port Coquitlam, Vancouver, Canada	Cable laying, autonomy, and communications	*ARCS*, Deep Ocean Logging Platform with Hydrographic Instrumentation and Navigation, *Theseus*, and *Aurora*
Japan Marine Science and Technology Center Marine Technology Department Yokosuka	Long-distance inertial navigation	Long-distance AUV
KDD Marine Engineering Laboratory Tokyo, Japan	Vision, cable tracking, and communications	*Aqua Explorer 2* and *Aqua Explorer 1000*
KISS Institute for Practical Robotics Norman, Oklahoma		*Dinky Robot in Pool*
University of Louisiana Apparel Computer Integrated Manufacturing Center Lafayette	Autonomous vehicle for underwater exploration	Phantom S2
Maridan Horsholm, Denmark	Design and manufacturing of AUVs	*Maridan*
Massachusetts Institute of Technology (MIT) AUV Laboratory at MIT Sea Grant Cambridge	Small, high-performance vehicles; nonacoustic sensors; energy management; docking; adaptive sampling; multiple vehicle operations; coastal modeling; object mapping; and under-ice, autonomous ocean sampling	*Odyssey II B*, Composite Endoskeleton Testbed Untethered Underwater Vehicle System, and *Altex*

Institution	Major Topics	Vehicles
Monterey Bay Aquarium Research Institute Moss Landing, California		*Dorado*
John C. Stennis Space Center Naval Oceanographic Office AUV Program Mississippi	AUVs	*Seahorse*
Naval Postgraduate School Center for AUV Research Monterey, California	Shallow-water applications	*Phoenix*
National Research Council of Canada Institute for Marine Dynamics Ottawa, Ontario	Canadian Self-Contained Off-the-Shelf Underwater Testbed (C-SCOUT)	C-SCOUT
Memorial University of Newfoundland Ocean Engineering Research Centre St. John's, Canada	C-SCOUT	C-SCOUT
Norwegian Underwater Intervention Bergen, Norway	Route and area surveys, search, and logging	*Hugin*
University of Port Laboratory of Systems and Subaqueous Technology Portugal	Autonomous and remote vehicles and control	*Isurus* and remote operated vehicles
Russian Academy of Sciences Institute of Marine Technology Problems Moscow	Solar-powered AUVs	
Sias Patterson Incorporated Gloucester Point, Virginia	AUVs	*Fetch2*

Institution	Major Topics	Vehicles
Simon Fraser University Underwater Research Laboratory Burnaby, British Columbia, Canada	Underwater acoustics, light-seeking AUVs, and autonomous sampling	*Purl, Purl II*
Southampton Oceanography Centre Ocean Engineering Division United Kingdom	Autonomous sampling and long-range missions	*Autosub*
University of Southampton Image, Speech, and Intelligent Systems Highfield, United Kingdom		*Neptune*
University of South Florida Center for Ocean Technology St. Petersburg	Sensors (optical, chemical, and acoustical) and seafloor classification	*BCAP*
Stanford University Aerospace Robotics Laboratory California	Dynamics, control, high-level command interface, and autonomy	Ocean Technology Testbed for Engineering Research
University of Sydney Australian Centre for Field Robotics	Position and attitude estimation and control	*Oberon*
Texas A&M University AUV Laboratory College Station		
Tokai University Kato Underwater Robotics Lab Shizuoka, Japan	Control, docking, and cable inspection	*Aqua Explorer 2* and *Aqua Explorer 1000*
University of Tokyo Ura Lab Japan	Autonomy, learning, long-range operations, and gliding vehicles	R1 (long-range autonomous operation), *Albac, Twin-Burger 2*, and *Manta-Ceresia*

Institution	Major Topics	Vehicles
Woods Hole Oceanographic Institution Deep Submergence Laboratory Massachusetts	Long-term seafloor monitoring, All kinds of marine operations	*Autonomous Benthic Explorer*, *Jason-Medea*, and *Remus*

SOURCE: Data from Institute of Electrical and Electronic Engineers, http://www.transit-port.net/Lists/AUVs.Org.html

Appendix D

Jason II and the New HOV Estimated Subsystem Weights and Costs

	Jason II		New HOV	
Subsystem	Weight (lb)	Cost (dollars)	Estimated Weight (lb)	Cost (dollars) Estimated
Structural				
Frame (including skids and tool sled)	560	27,000	1,900	95,000
Fairings			1,450	72,500
Hull (personnel sphere)			11,420	2,000,000
Pressure cases for electronics	850	58,000	850	58,000
Cabling and penetrators	83	28,000	160	56,000
Ballasting				
Variable ballast (hard tanks, pumps and motors, plumbing)			3,220	644,000
Soft ballast (soft tanks, high pressure air spheres)			800	160,000
Flotation (syntactic foam)	3,150	260,000	6,540	550,000
Energy and power source	287	18,000	2,430	160,000
Emergency ascent weight			1,360	68,000
Trim system (hardware for battery movement)			100	20,000
Propulsion (thrusters, motor controllers, electrical housings)	640	78,000	950	115,000
Power, external (conditioning and distribution)	Included in pressure case	28,000	Included in pressure cases	28,000

Instrumentation				
Control and display instrumentation (internal)			297	60,000
Fiber-optic telemetry	Included in pressure case	85,000	Included in pressure case	85,000
Sensors, external (lights, cameras, pressure and temperature, lasers)	306	163,000	281	163,000
Control instrumentation (external)	Included in pressure case	73,000	Included in pressure case	73,000
Life support (CO_2 scrubber, CO_2 absorbent, O_2 tanks, emergency breathing masks)			240	50,000
Navigation (gyro, doppler velocity log, computer, transducers)	292	176,000	120	75,000
Communications (radio, underwater telephone)			43	10,000
Hydraulic power unit	312	68,000	280	65,000
Manipulation system	382	285,000	320	300,000
Sample storage containers	117	18,000	150	25,000
Science payload reserve	240	N/A	800—light cond. 400—heavy cond.	N/A
Main cable (10-km length)	N/A	280,000	N/A	N/A
Traction winch	N/A	302,000	N/A	N/A
Control vans	N/A	370,000	N/A	N/A
TOTAL	7,200	2,300,000	33,700	6,500,000

NOTE: The source for the cost of *Jason II* and the weights of *Jason II* is A. Bowen, Woods Hole Oceanographic Institution, Woods Hole, MA, written communication, 2003. The source for the new HOV weights is R. Brown, Woods Hole Oceanographic Institution, Woods Hole, MA, written communication, 2003. The estimated cost for the new HOV is extrapolated on a weight basis from the cost of similar hardware for *Jason II*. The $6.5 million estimated cost for the new HOV is for hardware only and does not include design assembly, certification, and testing. WHOI has estimated the total cost of the new HOV at double the hardware cost.